Chocolate as Medicine
A Quest over the Centuries

Chocolate as Medicine

A Quest over the Centuries

Philip K. Wilson, Ph.D.
Professor, Department of Humanities
Director, The Doctors Kienle Center for Humanistic Medicine
Penn State College of Medicine
Hershey, Pennsylvania, USA
E-mail: pwilson@psu.edu

W. Jeffrey Hurst, Ph.D.
Principal Scientist
Hershey Foods Technical Center
Hershey, Pennsylvania, USA
E-mail: whurst@hersheys.com

RSC Publishing

ISBN: 978-1-84973-411-0

A catalogue record for this book is available from the British Library

Published by The Royal Society of Chemistry,
Thomas Graham House, Science Park, Milton Road,
Cambridge CB4 0WF, UK

Registered Charity Number 207890

For further information see our web site at www.rsc.org

Printed in the United Kingdom by Henry Ling Limited, Dorchester,
DT1 1HD, UK

For Janice and Deborah

Preface

The authors have held long-standing personal and professional interests in this topic. Wilson has led "symposia" for undergraduate students on this book's theme in classes taught at the University of Hawai'i at Mānoa and at Lehigh University. Professors at these institutions including Kerri Inglis, Beth Dolan and Monica Najar enthusiastically endorsed a further fleshing out of ideas that has resulted in this volume. Wilson has shared his research into chocolate's medical history in a number of presentations, beginning with "Good Taste in Therapeutic Choice: Our 'Addictive' Search for Chocolate's Medicinal Benefits" at the 2009 gathering of the Southern Association for the History of Medicine & Science in Birmingham, Alabama – a presentation at which chocolates were liberally distributed in an "airborne" manner that still lingers in the memories of many attendees.

Hurst traveled across the "one-company town" of Hershey to the College of Medicine where, in 2011, he offered a presentation on "Cocoa: From Ethnobotany to Translational Medicine" as part of the Hershey Society for the History of Medicine lecture series sponsored by The Doctors Kienle Center for Humanistic Medicine. Hurst has also spoken on "Chocolate: The Myth (Archeology and Ancient Medical Uses)" and "Chocolate: The Science (Advances in Science and Health)" at the Mt. Gretna, Pennsylvania community library. In 2010, Hurst attended the inaugural gathering of the International Society of Chocolate and Cocoa in Medicine

Chocolate as Medicine: A Quest over the Centuries
Philip K. Wilson and W. Jeffrey Hurst

Published by the Royal Society of Chemistry, www.rsc.org

(ISCHOM) held in Perugia, Italy. There, a group representing independent laboratories, academic departments and commercial confectionery companies from across Europe and the United States established an international society focused on the applications of chocolate in medicine. ISCHOM aims to interdisciplinarily bring together individuals and groups to promote future scientifically substantiated healthy uses of chocolate products in human dietary and medical ways that benefit the world's public.

Wilson and Hurst wish to specially recognize Dr Brian Puskas who, while a medical student at Penn State University's College of Medicine, initially brought them together through his own project on the "Cultural and Medical Uses of Chocolate in History", a work under Wilson's supervision that received the 2003 K. Danner Clouser Research Endowment Award from the College of Medicine's Humanities Department. Puskas shared his research at both the College of Medicine and the Hershey Company.

Interest in chocolate's potential medical benefits is booming across the globe. One recent work along these lines is Ronald Ross Watson, Victor R. Preedy and Sherma Zibadi's edited 40-chapter work on *Chocolate in Health and Nutrition* (2013).[1] A complementary publication, Rodolfo Paoletti, Andrea Poli, Ario Conti and Francesco Visioli's edited 12-chapter work *Chocolate and Health* appeared in 2012, in which Wilson provided the introductory historical chapter, "Chocolate as Medicine: A Changing Framework of Evidence Throughout History".[2] That chapter, together with Wilson's *Lancet* article on "Centuries of Seeking Chocolate's Medicinal Benefits" (2010), provided something of a précis for this more elaborate volume, the first book-length exploration of this particular historical quest.[3]

As in any substantial endeavor, the words on the page owe their origin to much more than the authors' handiwork. Most critically, a group of librarians and support staff from the authors' respective institutions have provided the professional expertise necessary for the soundness of this volume. From the beginning, Esther Y. Dell at the George T. Harrell Health Sciences Library at Penn State Hershey has been a stalwart supporter in helping to obtain needed material through interlibrary loan and via various electronic databases. Additional assistance from Penn State Hershey has been provided over the years by Deanne Bailey, Jeanne Brandt, Lori Coover, Ken Smith, Deb Tomazin, June Watson and Kathleen

Zamietra. Many staff member of Penn State's Pattee and Paterno Libraries on the University Park campus have also aided the research underlying this publication. Within the Informational Analysis Center (IAC) at the Hershey Company Technical Center, Jennifer R. Cessna, Colleen T. Shannon and especially Anna Venturella have extended every possible professional reference service that has allowed the research to be completed in a timely manner. Though the IAC is within Hurst's working environment, the Hershey Company generously provided Wilson temporary VIP status to use this important resource on an "as needed" basis, a courtesy that was gratefully appreciated. To work within the environment of an informational center solely dedicated to chocolate while the aroma of that product frequently permeated that space and having chocolate samples available *au discrétion* throughout the building is both a researcher's heavenly joy as well as a devilish temptation.

The entire Hershey, Pennsylvania community has provided support in its enthusiasm all throughout the research and writing stages of this volume. So many within the chocolate-rich heritage of Hershey have family and even direct connections back to the working world of Milton S. Hershey, Katherine Sweeney Hershey, and various company administrators, managers and supervisors who have unfailingly promoted this one company's connectedness with its community in a manner that so strikingly reflects Americana. Opportunities to delve into the Hershey Community Archives' advertisements of years past as well as into other pertinent historical matters were greatly supported by the Archives Director, Pamela C. Whitenack, as well as Tammy L. Hamilton, Archivist, and Kywin Zabolotny, Archives Assistant. Executive Director of the M.S. Hershey Foundation, Don Papson's support in the use of Archive illustrations for this work is much appreciated.

We also wish to thank our respective colleagues within the Hershey Center for Health & Nutrition and the Analytical Research and Services group as well as those in the Humanities Department of Penn State Hershey's College of Medicine for their support and encouragement. Special thanks are extended to Dr Daniel Azzara for his guidance and commitment to the Hershey Center for Health & Nutrition as well as to Dr David A Stuart for his leadership of this Center, especially in its formative years. The support, guidance and encouragement of four successive Humanities Department Chairs, Dr J.O. Ballard, Dr David J. Hufford, Dr John E. Neely, and Dr Daniel Shapiro are also much appreciated. We are grateful as well to Dr Richard Siderits who provided us with an electronic transcript of William Hughes' writing on chocolate, to Dr Larry Kienle for drawing our attention to chocolate and Colonial Williamsburg, to Dr Stephen Crozier for his valuable assistance with information included in Chapter 6, and to Dr Danny George for providing us with insights into the use of chocolate in dementia patient care. Scores of other individuals provided us with anecdotal accounts of their own positive experiences with chocolate, and we appreciated hearing each and every account.

Many individuals in the United Kingdom have also helped perpetuate the progress of this volume, most particularly the retired gastroenterologist and medical historian, Dr Denis Gibbs whose working library and hospitality provided together with Mrs Rachel Gibbs made for pleasurable work on repeated excursions to the Oxford environs. Lichfield physician Dr Chris Lockwood has routinely drawn our attention to work on chocolate as it has appeared in the British medical literature in ways similar to that of Penn State Hershey's Dr George Henning regarding United States medical works. The 2010 exhibition at the Royal College of Physicians in London on "Sir Hans Sloane: Discovery, Travels and Chocolate" provided sumptuous intellectual and gustatory appreciation of this icon's contribution to our knowledge of chocolate as medicine on the occasion of the 350th anniversary of Sloane's birth.

It is with profound gratitude that we acknowledged our deepest recognition to our respective spouses, Janice Wilson and Deborah Hurst. Without their enduring support – or their "chocoholic" tendencies – this work truly would not have held such special significance for its authors. Their critical readings of earlier drafts, their patient assistance at every stage, and Janice's use of her diligent librarianship in compiling our bibliography and index reflects the love that has so frequently embraced so much about chocolate throughout history. It is to these stalwart companions in our collaborative chocolate quest that we respectfully, humbly and lovingly dedicate this work.

It is the authors' great hope that this work is but a beginning in securing more regular collaborative efforts between individuals, centers and departmental groups at the Hershey Company and those at Penn State Hershey. Rather than each being perceived as "the other game in town", much mutual benefit in terms of biomedicine, public health, diet and business will surely rise from a more dedicated mixing of the key ingredients that each of these institutions holds such that the whole generated by such ventures may truly be greater than the sum of the parts.

P.K.W. & W.J.H.

REFERENCES

1. Ronald Ross Watson, Victor R. Preedy, and Sherma Zibadi, eds. *Chocolate in Health and Nutrition*, Humana Press, New York, 2013.
2. Philip K. Wilson, "Chocolate as Medicine: A Changing Framework of Evidence throughout History", in eds. Rodolfo Paoletti, Andrea Poli, Ario Conti, and Francesco Visioli, *Chocolate and Health*, Springer-Verlag Italia, Milan, 2012, pp. 1–15.
3. Philip K. Wilson, "Centuries of seeking chocolate's medicinal benefits", *Lancet*, 2012, **376**, pp. 158–159.

Contents

Chocolate as Medicine: A Quest over the Centuries
Philip K. Wilson and W. Jeffrey Hurst

Published by the Royal Society of Chemistry, www.rsc.org

List of Illustrations

Chocolate as Medicine: A Quest over the Centuries
Philip K. Wilson and W. Jeffrey Hurst

Published by the Royal Society of Chemistry, www.rsc.org

Chocolate as Medicine: An Introduction

In 1753, the noted nosologist, Carl Linnaeus, named the "Chocolate Tree" as *Theobroma cacao* – "Food of the Gods". Joanne Harris emphasized this exotic substance's erotic sensations in her 1999 fiction debut *Chocolat* as did Laura Esquivel in her premiere novel, *Como agua para Chocolate* (*Like Water for Chocolate,* 1989). Though the indefatigable private investigator Sherlock Holmes reputedly received some of his driving force from regularly injecting a 7% solution of cocaine, his energy was also boosted by the chocolate that he routinely consumed at breakfast. Gourmands and common consumers alike have formed subculture followings around their devotion to chocolate. Chocolate promotion and admiration societies have appeared in major world centers across the globe.

But is chocolate more than a confectionary delicacy? For millennia, healers have touted its myriad medicinal, albeit somewhat mystical, abilities. In 1906, it was argued that once the chocolate bar "finds a place in every pocket and home ... nerve specialists ... will lose their occupations".[1] By the 1950s, the chocolate that had once been used as a drug, a food and as a source of currency was being marketed merely as a pleasure-filled sugar snack. Half a century later, the craving to carve out chocolate's healthy, medicinal qualities resurged.

Chocolate as Medicine: A Quest over the Centuries
Philip K. Wilson and W. Jeffrey Hurst

Published by the Royal Society of Chemistry, www.rsc.org

> I haven't had this much good news since the early [19]70s when I learned I had passed all of the math requisites for my college degree. First it was the study that found napping was good for us and now it's the news that cocoa may boost brain function and delay decline as we age. That's right, two of my favorite things which previously had gotten bad raps, have now been determined to be good for me.
>
> Lou Ann Thomas, *Active Life* (2007)[2]

Few natural products have been purported to effectively treat such a wide variety of disorders as has chocolate.[3] Fortunately, allergies and side effects to chocolate are reportedly rare. Such claims readily lead to queries as to whether chocolate is, indeed, one of nature's greatest curative and restorative agents. When questioned, "What is Health?", the renowned 19th-century French gastronome, historian, philosopher and magistrate Jean Anthelme Brillat-Savarin replied, "It is Chocolate!".[4] Similar claims have been promoted in many chocolate company ads for decades (Figure I.1).

Long before the 21st century, chocolate's aphrodisiac aspect was purported. Chocolate "flatters you for awhile, it warms you for an instant; then all of a sudden, it kindles a mortal fever in you".[5]

> 'Twill make Old Women Young and Fresh
> Create New-Motions of the Flesh,
> And cause them long for you know what,
> If they but Tast[e] of Chocolate.
>
> Antonio Colmenero de Ledesma,
> *Chocolate; or, An Indian Drinke* (1652)[6]

London's Dr Henry Munday, according to M.L. Lémery's *Treatise on All Sorts of Foods . . . also of Drinkables* (1745), noted a patient "in a miserable condition" who, after "supping of Chocolate ... [was] recovered in a short Time; but what is more extraordinary is, that his Wife in Complacency to her husband, having also accustomed herself to sup Chocolate with him, bore afterwards several Children, though she was looked upon before not capable of having any".[7] Giacomo Casanova reputedly preferred chocolate to champagne to induce his seductive passions.

Figure I.1 "Hershey's For Health" from "The Story of Chocolate and Cocoa", Hershey's Company Promotional Booklet (1926).
(Courtesy of Hershey Community Archives, Hershey, Pennsylvania, USA).

More recently, the amorous properties ascribed to the Mars Company's green M&M'S have provoked a powerful urban legend. By the late 1970s, widespread belief in reputed aphrodisiacal powers found only in the green variety of M&M'S had taken hold. Though both the manufacturer and pop culture experts emphatically declared that the candies contained only chocolate, preservatives and coloring agents, the product's manufacturer Mars Snackfood U.S. later capitalised on this folklore when, on 16 January 2008, they issued a press release "proclaiming green the new color of love this Valentine's Day as the brand celebrates the myths, rumors and innuendo surrounding green M&M'S® Chocolate Candies". The "flirtatious, alluring, and confident Ms. Green M&M'S® brand character" appeared on the packages as did the disclaimer stating, "Consumption of The Green Ones® may result in elevated romance levels. If you experience this effect, contact your significant other immediately".[8]

Theobroma cacao is but one of the 22 species of the *Theobroma* plant. Cacao, originating from the Americas, was one of three imports (the others being coffee from Africa by 1544 and tea from Asia by 1610) whose active ingredients and reputations as medicines have been widely touted for centuries (Figure I.2). *Theobroma cacao* is the product of the "Chocolate Tree", a small, 4–8-m tall evergreen that grows naturally in the hot and rainy climates within

Figure I.2 Vignette of Coffee, Tea and Chocolate from Phillippe Sylvestre Dufour's *Traitez Nouveaux et Curieux du Café, du Thé et du Chocolat* (1688).
(Courtesy of Hershey Community Archives, Hershey, Pennsylvania, USA).

Figure I.3 Cacao Pods on the Chocolate Tree. From *The Story of Chocolate and Cocoa* (1926).
(Courtesy of Hershey Community Archives, Hershey, Pennsylvania, USA).

20 degrees north or south of the equator. These trees are typically intercropped with dense foliage plants including banana trees, plantains, lemon trees, coconut trees and the "Madre de Cacao", *Gliricidia sepiucum*, in order to provide the shade required for their optimal growth (Figure I.3).[9]

The cacao pods from which chocolate is derived grow straight out from the tree trunks rather than on stems. When young, they are green to red violet in color, and when ripe, yellow or orange. The pods vary somewhat in shape, some "resemble a small rugby ball, others a cucumber or a large gourd".[10] The tree's 50–100 pods per year each produce between 20–40 acrid and bitter seeds that are immersed in a white colored, sticky and bittersweet acidy pulp (Figures I.4, I.5, I.6 and I.7).

Three varieties of *Theobroma cacao* are harvested: *Crillo*, the indigenous tree once cultivated by the Maya and Aztecs that produce what are still revered as the prised beans; *Forastero*, the high

Cacao pods are cut off the high branches with a sharp knife fitted to the end of a long pole

Figure I.4 From *The Story of Chocolate and Cocoa* (1926). (Courtesy of Hershey Community Archives, Hershey, Pennsylvania, USA).

A sharp cut with the cutlass, a neat turn of the wrist
and the top half of the cacao pod is off

Figure I.5 From *The Story of Chocolate and Cocoa* (1926). (Courtesy of Hershey Community Archives, Hershey, Pennsylvania, USA).

yielding tree that originated from the upper Amazon region and that now, from Brazil and Africa, produces most of the beans used, especially in the mass production of chocolate; and *Trinitario*, the tree, though symbolically "sent from Heaven" as per its name, originated in Trinidad as a crossbreed between the other two varieties and whose beans are primarily used for blending (Figure I.8).[11]

Accounts of the labor-intensive harvesting of the Chocolate Tree appear in a wide range of works from *Harper's Monthly* journalist Henry Woodd Nevinson's *A Modern Slavery* (1906) to Jorge Amado's novel, *Cacáu* (1933) to Lowell J. Satre's *Chocolate on Trial: Slavery, Politics and the Ethics of Business* (2005) to Carol Off's *Bitter Chocolate: Investigating the Dark Side of the World's Most Seductive Sweet* (2008) to Órla Ryan's *Chocolate Nations: Living and Dying for Cocoa in West Africa* (2011) to Catherine Higgs' *Chocolate Islands: Cocoa, Slavery, and Colonial America* (2012). Such writings remind us that despite any healthfulness

After the cacao beans have been scooped out of the pods they are heaped on banana leaves or cocoanut matting

Figure I.6 From *The Story of Chocolate and Cocoa* (1926). (Courtesy of Hershey Community Archives, Hershey, Pennsylvania, USA).

attributed to chocolate, the growth, harvesting and processing of cacao beans has long produced significant hardships and diseases in cacao fieldworkers. Pressure from the Fair Trade Federation has at least begun to ameliorate the substandard labor and health conditions that have endured on many cacao plantations.

By the 19th century, chocolate was being consumed as a beverage for its reputedly wholesome, healthy and healing qualities. The corporate futures of several leading chocolate manufacturers grew out of this connection between their product and health. Philippe Suchard, for example, had first encountered chocolate when his ailing mother sent him to an apothecary in Neuchâtel, Switzerland to secure a pound of that product to incorporate into a tonic she was taking to improve her health. Stymied that a pound of chocolate equaled a "day's pay for a workman", this concern remained fervent in Suchard's mind until he was finally able, in 1826, to start producing a less costly yet equally healthy product for Swiss consumers.[12]

The drying of the cacao beans is done by spreading them
in a thin layer in the sun for at least three days

Figure I.7 From *The Story of Chocolate and Cocoa* (1926). (Courtesy of Hershey Community Archives, Hershey, Pennsylvania, USA).

Reports of chocolate's reputed medical benefits cross multiple areas of marketing. Chocolatiers as well as machine manufacturers have advertised such claims. F. Allan and Sons, for example, displayed their machinery for manufacturing chocolate and cocoa at the 1884 International Health Exhibition in London. More recently, as Susan Heller Anderson noted in her 1980 *New York Times* article, "Making Chocolates in the Artisan's Way", chocolate remains considered by many as "an aphrodisiac, a digestive, a soporific, a tonic, even a cure for certain intestinal afflictions".[13] Ian Knight, President of Knight International in Chicago, recognised the amazing staying power of chocolate as medicine in the introduction of his important edited volume, *Chocolate and Cocoa: Nutrition and Health* (1999). There, he acknowledged both historical and modern perspectives in claiming that although "historical texts may have lacked the sophistication of today's scientific methodology, they recorded benefits known at the time".[14]

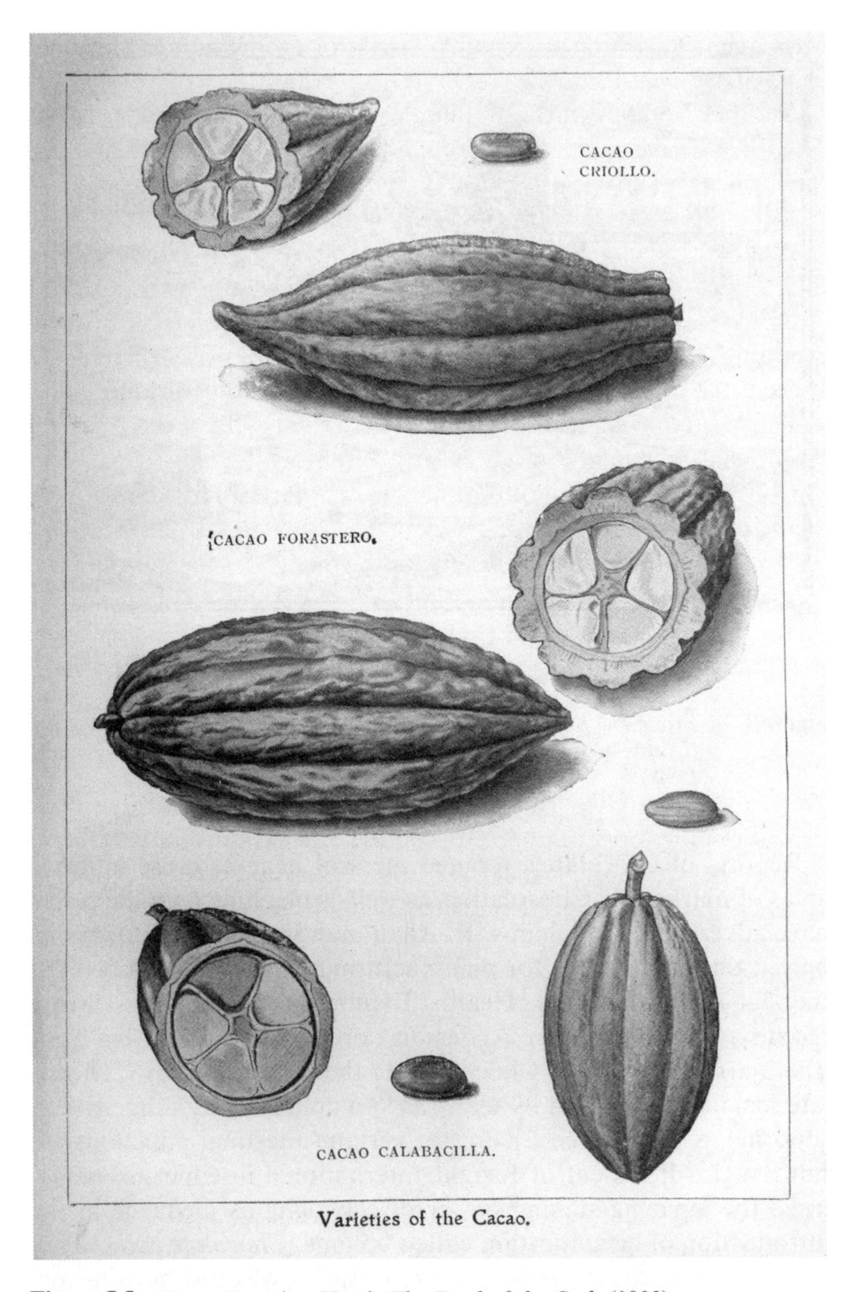

Figure I.8 From Brandon Head, *The Food of the Gods* (1903). (Courtesy of Hershey Community Archives, Hershey, Pennsylvania, USA).

Moreover, according to Knight, "most of their findings still hold some merit even today".[15]

Another popular author, Julie Pech, known as "The Chocolate Therapist™" has created a lucrative career based on her belief that this "once guilty pleasure has turned into the latest necessity of health".[16] Fortunately, this necessity is no longer only a high priced specialty product but rather one available to many people across the globe.[17] Still, chocolate's curious mystique prevails. For although it has become "democratized" and is touted to be "within everyone's budget", chocolate remains a highly prised product.[18] Indeed, the complexities of chocolate continue to confound us.

This volume explores evidence underlying the long-standing belief that chocolate (and cacao) offer health, medicinal and nutritional benefit. The authors begin with an abbreviated survey of the importance of seeking "Evidence" throughout medical history. Chocolate's various uses, including medical, are then explored in the Mesoamerican cultures that first experienced the power of this substance. The transport of this sylvan product to Europe and its transmogrification into a purely medical agent is discussed. Reports of chocolate's medical benefits are traced throughout Europe and then back across the Atlantic into Colonial North America. Further complexities of chocolate and its history become even more apparent when its seemingly duplicitous nature as both medicine and food of the 19th century is examined. In the final chapter before the epilogue, the authors focus upon more recent bioscientific experimental suggestions as the kind of evidence currently used to support chocolate's reputed efficacy as a medicinal and nutritional substance. This hitherto historical work then shifts its attention to the future, adding closing thoughts about several areas where chocolate's potential medical benefits are likely to be further investigated.

REFERENCES

1. "The increasing use of cocoa and chocolate in America", *Confectioners' Journal*, 1906, **32**, p. 72.
2. Lou Ann Thomas, *Active Life: For People 50-Plus On The Move*, 5 April 2007, p. 4.
3. See Appendix 1 for the most complete list to date of the disorders historically found to be treatable by chocolate.

4. Matty Chiva, "Cultural and Psychological Approaches to the Consumption of Chocolate", in *Chocolate and Cocoa: Health and Nutrition*, ed. Ian Knight, Blackwell Science, Oxford, England, 1999, pp. 321–338, p. 325.
5. Marquise de Sévigné (1671), as cited by John F. Mariani, "Sweet talk and chocolate", *MD*, 1993, **37**, pp. 89–91, p. 91.
6. Antonio Colmenero de Ledesma, *Chocolate; or, An Indian Drink*, J. Dakins, London, 1652.
7. Referring to Munday's 1685 *Opera Omnia Medico-Physica de Aëre Vitali, Esculentis et Potulentis cum Appendices de Paregis in Victu at Chocolatu, Thea, Caffea, Tobacco*.
8. "M&M'S® Chocolate Candies Go Green Just In Time For Valentine's Day". http://multivu.prnewswire.com/mnr/mars/31278/, accessed 28 April 2012.
9. Preparing chocolate from pod to product has been reviewed in a number of relatively recent works including Carol Ann Rinzler, *The Book of Chocolate*, St. Martin's Press, New York, 1977; Gordon Young, "Chocolate: Food of the gods", *National Geographic*, 1984, **166**, pp. 664–687; Ruth Mehrtens Galvin, "Sybaritic to some, sinful to others, but how sweet it is!", *Smithsonian*, 1986, **16**, pp. 54–64; Guy Mossu, *Cocoa*, Macmillan Press, London, 1992; Stephen T. Beckett, *The Science of Chocolate*, Royal Society of Chemistry, Cambridge, England, 2000; and Mericel E. Presilla, *The New Taste of Chocolate: A Cultural and Natural History of Cacao with Recipes*, Ten Speed Press, Berkeley and Toronto, 2001.
10. According to John Feltwell, "Cacao Plantations", in Nathalie Bailleux, Hervé Bizeul, John Feltwell, *et al.*, *The Book of Chocolate*, Flammarion, Paris and New York, 1996, pp. 17–58, p. 30.
11. William Gervase Clarence-Smith, ed., *Cocoa Pioneer Fronts since 1800: The Role of Smallholders, Planters and Merchants*, Macmillan Press, Houndmills and London, 1996, and William Gervase Clarence-Smith, *Cocoa and Chocolate, 1765–1914*, Routledge, London and New York, 2000 provide historical overviews of global chocolate production and consumption from a comparative economic history perspective.
12. Marcia Morton, *Chocolate: An Illustrated History*, Crown, New York, 1986, p. 27. Suchard had opened a confectionary shop in Neuchâtel, Switzerland in 1824 and, in 1826, he

founded a chocolate factory in Neuchâtel's Serrières neighborhood.

13. Anderson 17 December 1980, as cited by Linda K. Fuller, *Chocolate Fads, Folklore & Fantasies: 1,000+ Chunks of Chocolate Information*, Haworth Press, New York, 1994, p. 59.
14. Ian Knight, ed., *Chocolate and Cocoa: Health and Nutrition*, Blackwell Science, Oxford, England, 1999, p. 5.
15. For an overview of chocolate's relatively recent reputed medical benefits, see James N. Parker and Philip Parker, eds., *Chocolate: A Medical Dictionary, Bibliography and Annotated Research Guide to Internet References*, ICON Health Publications, San Diego, CA, 2003, http://www.netLibrary.com/urlapi.asp?action=summary&v=&bookid=99889, accessed 28 April 2012.
16. Julie Pech, *The Chocolate Therapist: Chocolate Remedies for a World of Ailments*, Trafford, Victoria, B.C., 2005, p. 91.
17. Such a broad statement needs a further bit of sensitivity and context. As Ambassador Ali Mchumo, Managing Director for the Common Fund of Commodities argued, cacao is a "commodity produced in the developing countries of the tropics and consumed mostly in the middle- and higher-income countries of the world's temperate zone", (A.B. Eskes and Y. Efron, eds., *Global Approaches to Cocoa Germplasm Utilization and Conservation*, CFC, Amsterdam, The Netherlands, 2006), preface. Allen Young shares a similar view in *The Chocolate Tree: A Natural History of Cacao*, University Press of Florida, Gainesville, 2007, p. x that today, "impoverished people in the tropics grow cacao, but chocolate, the processed product, is in large measure synonymous with Western affluence".
18. Matty Chiva, "Cultural and Psychological Approaches to the Consumption of Chocolate", in *Chocolate and Cocoa: Health and Nutrition*, ed. Ian Knight, Blackwell Science, Oxford, England, 1999, pp. 321–338, p. 335.

CHAPTER 1

Chocolate as Medicine: Seeking Evidence throughout History

> It is impossible to say how long the cultivation of cacao has existed, but it certainly goes back to very ancient times.
>
> C.J.J. van Hall, *Cacao* (1932)

Though "chocolate" is commonly referred to in a general sense, several distinctions should be made. Chocolate, itself, is the main processed byproduct of the cacao bean (or nib or cotyledon). Cacao, the species of the *Theobroma cacao* plant or "Chocolate Tree", is typically used in reference to the tree or pod or bean, whereas cocoa is used in reference to the powder made from the processed bean. Additionally, "chocolate" has carried different names in different cultures. For example, "chocolate" in Spanish, Portuguese and English languages was known as *chocotyl* to the Aztecs and *chocolatl* to Mexican Americans. In the Old World, the French called it *chocolat*, the Italians *cioccolata*, and the Germans *schokolade*. Russian languages refer to it as *shokoladno*. Except when keeping faithful to other uses in quotations, we will simply refer to this product as "chocolate" throughout this book.[1]

Histories of chocolate typically recount chronological discoveries regarding cacao seed (bean) products and the sequential improvements through which cacao has become used as medicine.[2] Among the earliest evidence for chocolate's medical use are

Chocolate as Medicine: A Quest over the Centuries
Philip K. Wilson and W. Jeffrey Hurst

Published by the Royal Society of Chemistry, www.rsc.org

the remaining iconographic works and fragments of Olmec, Maya, Zapotec, Mixtec and Aztec Art. Additional records from these eras are provided in groups of writings preserved under such names as the Florentine and Tuleda Aztec Codices as well as the Dresden and Madrid Mayan Codices. In recent decades, new forms of evidence have been uncovered in the remnants of *Theobroma cacao* found in the pottery and crockery of the Mokaya of Mesoamerica dating back to 1900 BC.[3]

Throughout human history, the importance and relative weight of oral tradition has been paramount, and indeed, has been the means by which people in earlier times generally learned of chocolate's potential health benefits. Various cultures speak of Quetzalcoatl as God of the air and, at least to the Aztecs, the patron saint of agriculture. On earth, this "Garden Prophet" lived in a beautiful sylvan grove where "students" of astronomy, medicine and agriculture would gather. It was there, so the story goes, that the special medicinal powers of the Chocolate Tree were discussed. The Aztec Emperor Moctezuma offered chocolate as his greatest gift to humankind as an apotheosis or glorification.[4] Based upon hearsay about such cultural uses, Carl Linnaeus gave the scientific name *Theobroma cacao* to the plant that provided the essential ingredient to this favorite drink of the Gods (Figure 1.1).

Looking at other forms of evidence, such as biogeography, we find that is was not so much the gathering of substances but, conversely, their spread across regions that secured chocolate's reputation. In this context, it was macaws and monkeys rather than Moctezuma who were responsible for naturally spreading the essence of cacao across a wide ecological area. By opening the cacao pods, devouring the luscious pulp, and leaving the bitter tasting beans where they dropped, it was these nonhumans who significantly expanded the terrain for future harvests. Though our primary focus is the human consumption and use of this natural product, we readily acknowledge this important nonhuman aspect of chocolate's early natural history.

Scholars from myriad fields have come to appreciate the importance of seeking indigenous knowledge in order to better understand those cultures that, for quite some time, had been termed as "primitive" and the "other". The Interinstitutional Consortium for Indigenous Knowledge (ICIK) at Penn State University in State College, Pennsylvania has become a leading

Figure 1.1 An Indigenous American Surrounded by a Chocolate Drinking Cup, a *Molinet*, and a Chocolate Pot, all atop an Image of the Cacao Pod. Frontispiece from Phillippe Sylvestre Dufour's *Traitez Nouveaux et Curieux du Café, du Thé et du Chocolat* (1688). (Courtesy of Hershey Community Archives, Hershey, Pennsylvania, USA).

program in academe that studies indigenous knowledge as a unique focus. According to ICIK, indigenous knowledge is

> an emerging area of study that focuses on the ways of knowing, seeing, and thinking that are passed down orally from generation to generation, and which reflect thousands of years of experimentation and innovation in everything from agriculture, animal husbandry and child rearing practices to education; and from medicine to natural resource management. These ways of knowing are particularly important in the era of globalization, a time in which indigenous knowledge as intellectual property is taking new significance in the search for answers to many of the world's most vexing problems – disease, famine, ethnic conflict, and poverty. Indigenous knowledge has value, not only for the culture in which it develops, but also for scientists and planners seeking solutions to community problems.... [including] health, agriculture, education, and the environment, both in developed and in developing countries.[5]

In terms of chocolate, "indigenous" refers to the equatorial Americas in the lowland forests of the Amazon–Orinoco basin flood plain. From there, colonisation boosted the cultivation of cacao beans into the realm of a major industry in tropical equatorial regions across the globe. Despite decades of dedicated interest in improving the harvest yield of cacao in these regions, chocolate's indigenous history has only recently garnered significant academic interest. Best among these works is Cameron L. McNeil's edited volume, *Chocolate in Mesoamerica: A Cultural History of Cacao* (2007). A complete indigenous history of chocolate as medicine in Mesoamerica awaits an author.

Work in this volume focuses upon history drawn from various types of evidence, ranging from tradition to testimonials to travel narratives as well as crockery, case studies and cookbooks. Covering more recent centuries, the power of advertising is analysed as another important source of surviving recorded evidence. The closing chapters focus more upon biomedical evidence. Still, as this work makes apparent, word-of-mouth (*i.e.*, a type of oral tradition sharing indigenous knowledge) has never really diminished as a key source of evidence in promoting chocolate's healing powers.

"Evidence" has become an increasingly important buzzword within the healing arts over the past few decades. For instance, this term appears in the subtitle of David Katz's *Nutrition in Clinical Practice: A Comprehensive Evidence-Based Manual for the Practitioner*, a work which includes a chapter on the "Health Effects of Chocolate". In many ways, keeping evidence as a central thread throughout our historical account has helped us retain a pertinent focus upon chocolate's potential therapeutic effects. To better appreciate this historical thread when applied to chocolate, a brief reflection upon the growth and meaning of "evidence" used to support general biomedical and health claims over the past few centuries is warranted.

1.1 VALUING MEDICAL "EVIDENCE" IN THE PAST AND PRESENT

In our era, the therapeutic efficacy of medical practices and remedies has been recast within the mold of evidence-based medicine (EBM). Beginning in the early 1990s, the then newly established clinical discipline of EBM referred to the "conscientious, explicit, and judicious use of current best evidence in making decisions about the care of individual patients".[6] Clinical expertise is combined with newly supported biomedical evidence obtained through systematic literature searches to ensure the delivery of the highest quality health care. EBM also incorporates a "thoughtful identification and compassionate use of individual patients' predicaments, rights and preferences in making clinical decisions about their care".[7] Including the patient in the decision-making process conforms with the late Cornell University internist, Eric Cassell's notable directive that to effectively relieve suffering, physicians must be ever mindful of a patient's entire personhood.[8]

Since the early 1990s, EBM has been adopted into medical school curricula, celebrated in medical manuals, deliberated in medical literature and featured at myriad medical conferences. The vast international consortium of clinicians and consumers known as the Cochrane Collaboration has provided systematic literature reviews as required by EBM methodology. Despite its popularity, arguments have surfaced claiming that EBM is either "old hat" or "impossible to practice".[9] At the heart of the matter lies the concern over acknowledging what has truly counted as "evidence" in different eras of medicine's heritage.

Around the time when chocolate was first introduced into European culture, "evidence" was reexamined within scientific and medical contexts. Sir Francis Bacon, best known at the time for his financial prowess as Lord Chancellor under the reign of England's James I, is credited with providing a new framework for science: the experimental method (Figure 1.2). If the purpose of science was, as he argued, to give humans mastery over nature, thereby extending both human knowledge and power, then the laws of nature must be better understood. Such understanding, so Bacon proclaimed in *Novum Organum* (1620), was attainable only after shifting scientific thought from deductive reasoning towards an inductive approach coupled with experimentation.

Bacon's inductive method of interpreting nature – which others later applied to chocolate – involved the assembly of a "sufficient, . . . accurate collection of instances", or evidence, gathered "with sagacity and recorded with Impartial plainness". After viewing it "in all possible lights, to be sure that no contradictory . . . [evidence] can be brought, some portion of useful truth", general law, or hypothesis could then be established.[10] He argued that natural philosophers who relied solely upon the authority of the past – which for all university graduates of his day was still the ancient logic

Figure 1.2 Sir Francis Bacon, from title page of David Mallet's *The Life of Francis Bacon, Lord Chancellor of England* (1740).

(or *Organon*) of Aristotle – failed to advance any new understanding of nature. Bacon advocated the experimental method as the most reliable manner to free science from the "paralysing dependence of previous students of nature on the rough and ready conceptual equipment of everyday observation".[11]

Like science, medicine had also long been practised according to ancient dogma. Hippocratic wisdom proffered guiding aphorisms of the medical art. Contrary to Aristotelian reasoning, Hippocratic diagnoses stemmed from developing a general hypothesis based upon carefully observing specific signs and symptoms. Yet according to Bacon, Hippocratic doctrine had become "more professed than labored" by the early 1600s. Subsequently, he queried how to ascertain which contemporary medical practices yielded the very best possible outcomes.

Bacon's concerns regarding medical practice stemmed from observing how a theory-based approach had come to prevail over a (Hippocratic) patient-oriented one. Adding support to Bacon's attempt to reinvigorate Hippocratic perspectives, one seventeenth-century practitioner, Thomas Sydenham – later dubbed the "English Hippocrates" – vehemently opposed theory-based medicine claiming, instead, that medicine should be practised by first objectively gathering signs and symptoms without prematurely speculating upon their significance. Then, only after distinguishing useful signs from red herrings could the true understanding of a disorder become realised. By restoring this Hippocratic inductively derived diagnosis, expected disease patterns could then be deduced. The London surgeon and physician, Daniel Turner echoed Sydenham in the following century, claiming that disease was not *a priori* predictable according natural laws. Physicians were not like natural philosophers who were free to apply rules "to Bodies inanimate, or putting simple Fluids into ... Balance ... [or] counting .... Pressures or Impulses". Rather, they were dealing with human lives. Physicians, he asserted, must not "sacrific[e] Men's Lives" for the sake of some "meer [sic] Hypothesis".[12]

Eighteenth-century medical practitioners frequently relied upon testimonial evidence to discern the efficacy of particular remedies. London physician James Jurin, for example, gathered testimonials in "good Baconian fashion", tabulated the results, and based his conclusions upon "matters of fact" that could be demonstrated numerically.[13] Through numerical representations, a patient's anonymity would be maintained, a focus on success would sidestep religious and ethical debate, and charges of quackery would be squelched. In conclusion, basing medical practices upon evidence derived from this numerical method made them appear "more philosophical and hence, legitimate".[14]

"Experience" and "experiment", two expressions that were synonymous in Romance languages, were also used interchangeably in discussing Bacon's vision of evidence-based healthcare. For Bacon, only "ordered experience" that was founded upon methodological investigation, measurable criteria and objectivity counted as "evidence", whereas "ordinary experience" based solely upon chance observation and subjectivity did not.[15] His suggestions for revolutionising the experiential and experimental basis of science were more formally embodied in the formation of London's Royal Society in 1660. This elite body, whose Fellows included the city's leading physicians and many prominent promoters of chocolate for health, undertook the task of critically appraising the current state of knowledge. Their motto, *nullius in verba* – upon the word or authority of no one – stressed the Society's reliance upon experiment and personal experience over preconceived theorisation. Moreover, regular interaction among the Fellows exemplified Bacon's description of a utopian "college of experience", where open discussion and collaboration between investigators were encouraged. This Society provided the first venue in which members with similar interests gathered to listen to reports of each other's experiences with various natural phenomena. The accounts in their publication, the *Philosophical Transactions*, were written from the viewpoint of the observer and, by convention, they contained details of the time, place and participants or witnesses of a particular experience. The elaborate narrative details in the reports were rhetorically constructed consistent with those of Royal Society Fellow and experimenter Robert Boyle, thereby giving them the "impression of verisimilitude" and compelling the Fellows to accept the reported details as "matters of fact".[16]

1.2 GATHERING NUMERICAL EVIDENCE

Medical decision making according to these standards became increasingly practised in the eighteenth century. Historian Ulrich Tröhler's critical examination of quantification in eighteenth-century therapies offers insight into "evidence" as used in medical practice of that period. Indeed, he focused upon contemporary efforts *To Improve the Evidence of Medicine* – a phrase gleaned from London physician, George Fordyce's 1793 publication. As an example, John Millar, a practitioner at the Westminster Dispensary, claimed in 1777 that "arithmetical calculation" provided "incontestable evidence" of one therapy's benefit over another. He deemed that individual case reports, no matter how "numerous and well attested" were "insufficient to support general conclusions". However, "by recording every case in a public and extensive practice, and comparing the success of various methods of cure with the unassisted efforts of nature", some "useful information may be obtained; and the dignity of the profession may be vindicated from vague declaration and groundless aspersions". Where "Mathematical reasoning can be had", Millar concluded, "it is [as] ... great [a] folly to make use of any other [method], as to grope for a thing in the dark, when you have a candle standing by you".[17]

By the turn of the 21st century, claims of chocolate's benefits increasingly became based upon expanding amounts of data. Relying upon these large data sets instead of merely upon one single practitioner's experience had also become a significant feature for some 18th-century practitioners. As Scottish naval surgeon Robert Robertson argued, "few practitioners in physic will have so many cases come under their care, ... and consequently ... few readers will ever have experimental authority to deny the validity" of particular remedies.[18] Still, small-scale experiments were viewed to be of some importance, as seen in the comparative clinical trials to cure scurvy. James Lind's success, described in 1753, of treating scorbutic (*i.e.*, scurvy suffering) sailors by giving them "two oranges and one lemon daily for six days" was based on a very small-scale trial. Comparative methods on a larger sampling of patients, including the sailors aboard James Cook's voyages, suggested that the Irish physician, David Macbride's recommendation of "wort" (*i.e.*, unfermented malt) in the diet appeared the most promising preventative. Still, Lind persisted with "clinical trials" for another twenty years, treating upwards of 400 scorbutic patients a day in England's Haslar Hospital during the Seven Years' War (1756–63). In 1772, his arithmetic persuasion won the day. As he reflected, a "work ... more perfect, and remedies more absolutely certain might perhaps have been expected from an inspection of several thousand scorbutic patients, from a perusal of every book published on the subject, and from an extensive correspondence with most parts of the world ... but, though a few partial facts and observations may for a little [time], flatter with hopes of great success, ... [an] enlarged experience must ever evince the fallacy of all positive assertions in the healing art".[19]

George Guthrie, a British surgeon active during the Peninsular War of Spanish Independence, did not trust any "theory or opinion of authors not supported by actual experience". Regarding the immediacy of amputation on the battlefield – a critical concern for all Napoleonic War surgeons – he recommended what we would recognise as a prospective comparison study. "It is not sufficient to perform twenty amputations on the field of battle, and contrast them with as many cases of amputation, done at a later period", he argued. The "twenty cases for delayed operation must be selected on the field of battle, and their result compared" some time later. Only through such comparison, he concluded, would "the value of the two modes ... be duly estimated".[20]

British healers of the 1700s promoted new types of evidence in their campaigns for medical reform. They demanded "adequate empirical evaluation of existing and [newly] proposed ... remedies". Doing so required "extensive, comparative trials, with results expressed by numbers". They "wished to base clinical medicine on elementary numerical analysis of compilations of observations made on distinct groups of patients". Their "observational analyses" of the efficacy of their treatments employed mortality statistics, administrative returns and personal case histories that included both "one clinician's experience with patients receiving a particular treatment, as well as compiled statistics describing the experience of several clinicians" in treating the same disorder.[21]

The quest for more practical information gave rise to hospital-based practices in the 1700s and the "Birth of the Clinic" in the early 1800s. In both types of institutions, the usefulness of evidence expanded from diagnosis to treatment. The "evidence" of particular treatments was increasingly represented in arithmetic ratios. For Pierre-Charles-Alexandre Louis, a leading hospital surgeon in the heavy chocolate consuming city of Paris, this numerical method – or medical arithmetic – became synonymous with scientific reasoning. The clinician reasoned by "employing aggregative thinking about a population of sick individuals rather than using pathological anatomy to observe disease in a particular individual". By emphasising "quantitative thinking at the level of the social group rather than on qualitative understanding at the level of the individual patient", Louis was "appeal[ing] to the authority of number to justify clinical judgment". With a triumph of the numerical method, Louis argued that "No treatise whatsoever will continue to be the sole development of an idea". Rather, once an "analysis of a[n] ... extensive series of exact detailed facts" furnishes answers to "all possible questions", therapeutics will "become a science".[22] Westminster Dispensary physician, Francis Bisset Hawkins made a similar claim in his 1829 work entitled *Elements of Medical Statistics*. There is "reason to believe that a careful cultivation of [statistics], in reference to the natural history of man in health and disease, would materially assist the completion of a philosophy of medicine .... Medical statistics affords the most convincing proofs of the efficacy of medicine".[23]

Credible sources abound suggesting that particular elements of what we regard as EBM existed in previous eras. Today, EBM practices rely upon internet abstracting services and international consortia such as the Cochrane Collaboration to systematically cull pertinent information from the ever-expanding warehouse of knowledge. However, the preparation of abridgements and abstracts of medical cases has been a practice for centuries, among the earliest being the 1745 abstract of key articles published in the Royal Society's *Philosophical Transactions*. Thomas Southwell's abridgement of medical breakthroughs presented before the Royal Academy of Science of Paris appeared in his *Medical Essays and Observations* in 1764.

Since the 1700s, abstracting key medical discoveries has escalated in parallel to the rise of medical periodical publishing. Many of these abstracts reflect the growing importance of surveying a series of multiple cases and analysing them with statistical methods. For younger physicians, then and now, whose catalog of personal experiences does not always meet their decision-making needs, relying upon abstracts has long been critical to their practices. But in contrast to earlier eras, modern technology has enabled health-care providers to fulfill what were only dreams in bygone eras to quickly cull the most pertinent information from ever-expanding databases of practices. Today, we have realised what the Irish physician, William Black argued in his *Arithmetic and Medical Analysis of the Diseases and Mortality of the Human Species* over two centuries ago, that an evidence-based medical arithmetic will be the dawn of a new era of medicine.

In sum, EBM is quintessentially history-based medicine. Historians work detective-like, piecing together information from past events, sifting through red herrings as they seek for the best evidence upon which to base their interpretation, diagnosis and treatment. Similar to developing persuasive and accurate historical accounts, EBM carefully critiques a wide range of information in order to distinguish credible from incredible sources. Like the reputable historian, the EBM-guided physician has, through assimilating sources and ascertaining their validity, developed solid reasons for pursuing particular treatment protocols. Today, EBM is most often directed toward securing the best treatment from a host of historically produced possibilities.

Though we have seen that some medical and surgical practitioners readily adopted the experimental method as a means of gaining evidence, many continued to rely, at least in part, upon anecdotes and individual case studies as guiding lights. Support for chocolate's therapeutic abilities has also been garnered from various forms of evidence, anecdotes being the chief form that substantiated its use as medicine once this product reached the Old World. Such anecdotes, like chocolate itself, initially gained root half a globe away.

REFERENCES

1. For further linguistic exploration of the Mixe-Zoque languages of Olmec traditions, see Karen Dakin and Søren Wichmann, "Cacao and chocolate: A Uto-Aztecan perspective", *Ancient Mesoamerica*, 2000, **11**, pp. 55–75.
2. Among the vast number of general works on chocolate which, to varying degrees, chronologically review the potential health claims are Albert Bourgaux, *Quatres Siècles d'histoire du Cacao et du Chocolat*, Office International du Cacao et du Chocolat, Brussels, 1935; Eula Long, *Chocolate: From Mayan to Modern*, Aladdin Books, New York, 1950; J. Eric S. Thompson, "Notes

on the Use of Cacao in Middle America", *Notes on Middle American Archaeology and Ethnology*, Carnegie Institution of Washington, Department of Archaeology, 1956, **128**, pp. 95–116; Norah Smaridge, *The World of Chocolate*, J. Messner, New York, 1969; Norman Kolpas, *The Chocolate Lovers' Companion*, The Felix Gluck Press, Twickenham, England, 1977; Marcia Morton, *Chocolate: An Illustrated History*, Crown, New York, 1986; Nikita Harwich, *Histoire du Chocolat*, Editions Desjonquères, Paris, 1992; Nathalie Bailleux, Hervé Bizeul, John Feltwell, *et al.*, *The Book of Chocolate*, Flammarion, Paris and New York, 1996; Sophie D. Coe and Michael D. Coe, *The True History of Chocolate*, Thames and Hudson, London, 1996; Christine McFadden and Christine France, *Chocolate: Cooking with the World's Best Ingredient*, Hermes House, New York, 2001; Mort Rosenblum, *Chocolate: A Bittersweet Saga of Dark and Light*, North Point Press, New York, 2005; Meredith L. Dreiss and Sharon Edgar Greenhill, *Chocolate: Pathway to the Gods*, University of Arizona Press, Tucson, 2008; and Phillip Minton, *Chocolate: Healthfood of the Gods: Unwrap the Secrets of Chocolate for Health, Beauty and Longevity*, 2011, available through Amazon.com and bn.com. See also the online Chocolate Research Portal, https://cocoaknow.ucdavis.edu/ChocolateResearch, accessed 29 April 2012.

3. W. Jeffrey Hurst, M. Stanley Tarka, Jr., G. Terry Powis, *et al.*, "Archaeology: Cacao usage by the earliest Maya civilization", *Nature*, 2002, **418**, pp. 289–290.
4. Philip Porter Gott and L. F. Van Houten, *All About Candy and Chocolate: A Comprehensive Study of the Candy and Chocolate Industries*, National Confectioners' Association of the United States, Chicago, 1958, p. 33.
5. See http://www.ed.psu.edu/icik, accessed 30 April 2012. Penn State scholars including Dr Audrey N. Maretski and Dr Ladislaus M. Semali worked to craft this definition in the 1990s.
6. D. L. Sackett, W. M. C. Rosenberg, J. A. M. Gray, *et al.*, "Evidence-based medicine: What it is and what it isn't", *BMJ*, 1996, **312**, pp. 71–72, p. 71. Philip K. Wilson, "Weighing medical evidence on a historical scale", *Hektoen International: A Journal of Medical Humanities*, 2011, **3**, www.hektoneinternational.org/Weighing-medical-evidence.html offers a condensed version of

this chapter. See also Philip K. Wilson, "Chocolate as Medicine: A Changing Framework of Evidence Throughout History", in *Chocolate and Health*, eds. Rodolfo Paoletti, Andrea Poli, Ario Conti and Francesco Visioli, Springer-Verlag Italia, Milan, 2012, pp. 1–15.

7. D. L. Sackett, W. M. C. Rosenberg, J. A. M. Gray, *et al.*, "Evidence-based medicine: What it is and what it isn't", *BMJ*, 1996, **312**, pp. 71–72, p. 71.
8. E. J. Cassel, "The nature of suffering and the goals of medicine", *New England Journal of Medicine,* 1982, **306**, pp. 639–645.
9. D. L. Sackett, W. M. C. Rosenberg, J. A. M. Gray, *et al.*, "Evidence-based medicine: What it is and what it isn't", *BMJ*, 1996, **312**, pp. 71–72, p. 72.
10. D. Mallet, *The Life of Francis Bacon, Lord Chancellor of England*, A. Millar, London, 1740, pp. 93–94.
11. A. Quinton, *Francis Bacon*, Oxford University Press, Oxford, 1980, p. 55. Wilson briefly recounted the Baconian influence upon revolutionising science in Philip K. Wilson, "Origins of science", *National Forum* (Journal of the National Honor Society, Phi Kappa Phi), 1996, **76**, pp. 39–43. Also reprinted in SIRS Renaissance electronic database, 1996.
12. Daniel Turner, *Art of Surgery*, vol. ii, Rivington, Lacy and Clarke, London, 1725, p. 75, and Daniel Turner, *Discourse Concerning Fevers*, Oake, London, 1727, pp. 28–29, 58. For a review of Turner's influence, see Philip K. Wilson, *Surgery, Skin & Syphilis: Daniel Turner's London (1667–1741)*, Wellcome Institute Series in the History of Medicine, Clio Medica **54**, Rodopi Press, Amsterdam and Atlanta, 1999.
13. A. A. Rusnock, "The Weight of Evidence and the Burden of Authority: Case Histories, Medical Statistics and Smallpox Inoculation", in *Medicine in the Enlightenment*, ed. R. Porter, Rodopi, Amsterdam, 1995, p. 305.
14. A. A. Rusnock, "The Weight of Evidence and the Burden of Authority: Case Histories, Medical Statistics and Smallpox Inoculation", in *Medicine in the Enlightenment*, ed. R. Porter, Rodopi, Amsterdam, 1995, p. 305.
15. U. Tröhler, *To Improve the Evidence of Medicine: The 18th Century British Origins of a Critical Approach*, Royal College of Physicians of Edinburgh, Edinburgh, 2001, pp. 1–2.

16. S. Shapin, "Pump and circumstance: Robert Boyle's literary technology", *Social Studies of Science*, 1984, **14**, pp. 487–494, p. 493.
17. J. Millar, *Observations on the Prevailing Diseases in Great Britain*, 2nd edn, Millar, London, 1798, p. 76.
18. R. Robertson, *Observations on the Jail, Hospital or Ship Fever*, Murray, London, 1783, p. 312.
19. J. Lind, *A Treatise on the Scurvy*, 3rd edn, Crowder, Wilson, Nicholls, Cadell, Becket, Pearch and Woodfall, London, 1772, pp. v–vi.
20. G. J. Guthrie, *On Gun-Shot Wounds of the Extremities*, Longman, Hurst, Rees, Orme and Browne, London, 1815, p. 39.
21. U. Tröhler, *To Improve the Evidence of Medicine: The 18th Century British Origins of a Critical Approach*, Royal College of Physicians of Edinburgh, Edinburgh, 2001, p. 115.
22. P.-C.-A. Louis, *Researches on the Effects of Bloodletting in Some Inflammatory Diseases, and On the Influence of Tartarized Antimony and Vessication in Pneumonitis*, translated by C.G. Putnam, Hillard, Gray, Boston, 1836, p. 70.
23. F. B. Hawkins, *Elements of Medical Statistics*, Longman, London, 1829, pp. 2–3.

CHAPTER 2

Chocolate and Healing in Pre-Columbian Mesoamerican Culture

Chocolate is an alimentary preparation of very ancient use in Mexico.

Andrew Ure, *Dictionary of Arts, Manufactures and Mines* (1839)

Before the development of chemical residue testing of pottery vessels, it was difficult to determine when pre-Columbian Mesoamericans first consumed cacao.

Cameron L. McNeil, *Chocolate in Mesoamerica: A Cultural History of Cacao* (2006)

In the 1940s while serving as an army surgeon in the Panama Canal Zone, B.H. Kean recorded his findings regarding hypertension among the Kuna Indians, descendants of the Mayans, who resided in the relatively isolated region of the San Blas archipelago off the southern end of Central America. Kean noted that the Kuna adults showed a remarkably low yet stable blood pressure.[1] Half a century later, Harvard physician Norman K. Hollenberg who, at the time, was researching potential genetic links to hypertension, discovered Kean's papers, whereupon he sought to investigate whether the Kuna Indians still experienced an

Chocolate as Medicine: A Quest over the Centuries
Philip K. Wilson and W. Jeffrey Hurst

Published by the Royal Society of Chemistry, www.rsc.org

extremely low incidence of hypertension. Turning his attention from nature to nurture, Hollenberg found an environmental effect in that the Kuna Indians who had left San Blas and resettled in Panama City exhibited higher rates of hypertension than did those who had remained on the archipelago. Among the key differences noted between these distinct groups of peoples was their respective dietary use of chocolate. The more traditional Kuna diet included natural or at most very lightly processed cacao drinks that they consumed five times daily, whereas the Kuna living in Panama City drank only a highly processed cocoa, though in lesser amounts.[2] This finding prompted an explosion of claims, though initially little substantiated, about ancient uses of unprocessed chocolate that might well benefit modern medical practices.

2.1 MESOAMERICAN CULTURE AND CHOCOLATE

Before turning to healing stories and myths surrounding the Kuna ancestors' use of chocolate in Mesoamerican culture, a brief overview of a few cultural groups and key individuals may be helpful.[3] The Mixe-Zoquean-speaking Olmec were among the first peoples who settled in the equatorial region of the Amazon–Orinoco basin of the Mexican Gulf Coast and the first to leave striking evidence of their culture and their settlement. They inhabited areas surrounding the present-day Mexican state of Tabasco between 1500 and 400 B.C. Planters within the Izapan culture – a group that bridged the traditional Olmec and the Classical Maya periods and covered regions reaching from the Yucatán peninsula to the Pacific Coastal plains of what is now Chiapas, Mexico and along the Pacific coast of Guatemala – were likely the first to harvest cacao.[4]

The Chocolate Tree appeared in the later Mayan creation myth, *The Popul Vuh* – the "Book of Counsel". According to Mayan legend, cacao originated during the reign of the third Mayan king, Hunahpú, and it was given to humans by the Sovereign Plumed Serpent God.[5] Yet cacao became most closely associated with another ruler, the invading Toltec leader, Topiltzin Quetzalcoatl who invaded the Mayan lands from the north, precipitating the Classic Maya Collapse and the beginning of what would become the Aztec (or Mexica) rule. Shortly after settling in his newly acquired territory in the Yucatán town of Chichén Iztá, Quetzalcoatl became deified, initiating his destiny within Aztec mythology.

According to the Franciscan monk, Bernardino de Sahagún's 1590 ethnographic and encyclopedic chronicles of life and culture in New Spain from 1529–1589, Quetzalcoatl soon came to possess "All of the riches of the world in gold and silver ... as well as an abundance of cacao trees of various colors".[6] The Toltec reign came to an end when Quetzalcoatl, in the form of a human ruler, fell from grace and was driven insane from a mysterious potion; ultimately banished, he fled the region on a raft supposedly composed of snakes. Before leaving, however, Quetzalcoatl – now as the Plumed Serpent God – promised to return, bringing along all the treasures of Paradise.

In 1519 – the very year that Mesoamerican astrologers had predicted Quetzalcoatl's return – during the reign of the last Aztec emperor Moctezuma II, a "vessel full of men whose armor shone in the sun like serpent's scales" and who were "crowned with plumed helmets" arrived on the shores of the West Indies (present day Veracruz).[7] Thinking the leader of these "gods" to be Quetzalcoatl, Moctezuma (himself, a hearty consumer of chocolate) surrendered his territory to the leader who, in reality, was Hernán Cortés and to his *conquistadores*. Like Moctezuma's warriors who were given chocolate drink (*xocolatl*) to maintain their endurance, Cortés soon claimed chocolate to be the "Divine Drink" which "builds up resistance and fights fatigue". He offered further testimonial, perhaps hearsay, evidence that, "a cup of this precious drink permits a man to walk all day without food" (Figure 2.1).[8]

During the Conquest that followed, and the sacking of Tenochtitlán (present-day Mexico City) in 1521, many insights into Aztec culture that might have been available for later generations were destroyed. Thus, much of what is believed about this culture is derived from accounts of the *conquistadores*, a few codices transcribed in the 20th century, decoded hieroglyphic-decorated vessels, stelae, and sculptured building façades, and more recent chemical analysis of archeological finds.[9] Among the latter, W. Jeffrey Hurst's chemical analysis has helped identify cacao residue in drinking vessels and cookery pots from Middle Preclassic/Formative Mesoamerica (600–300 B.C.) in Belize, and in Early Classic Mesoamerica (250–600 A.D.) vessels from Río Azul, Guatemala and Copan, Honduras (Figure 2.2).[10]

Figure 2.1 Hernán Cortés and Moctezuma, *The Story of Chocolate and Cocoa* (1934).
(Courtesy of Hershey Community Archives, Hershey, Pennsylvania, USA).

Traces of theobromine have been detected in Early Classical Mayan ceramic vessels dating from 460 to 480 A.D. that were unearthed from a tomb at the Mayan archeological site of Río Azul in northeastern Guatemala. Two of the hieroglyphs on one of the vessels are the syllabic spellings for the word, *ca-ca-wa*. The "position of these glyphs within the text [on this vessel] makes it virtually certain that they refer to the substance that the vessel was meant to contain".[11] The hieroglyphs themselves do not illuminate specific reasons why the Mayan peoples consumed chocolate; however, other anthropological evidence drawn from geographical locations throughout this region strongly suggests that rather than consuming cacao in the manner once thought predominant among these cultures, it was most likely used as the base for alcoholic beverages. Similar chemical analysis of residues taken from spouted ceramic vessels uncovered at an archeological site at Colha in what is now northern Belize suggests that cacao was prepared in beverage form at least as early as 600 B.C. during the Preclassic Maya period, thereby "pushing back the earliest chemical evidence of cacao use by some 1,000 years" (Figure 2.3).[12]

In both Mayan and later Aztec cultures, cacao-based beverages were served to a variety of different individuals in society,

Figure 2.2 Early Classic Mayan Vessel that Once Held Chocolate Drink, Found in Tomb at Río Azul, Guatemala.
(Used with Permission of Dr Grant D. Hall, Texas Tech University).

including male priest figures, high chiefs and distinguished warriors – all of whom had offered much to the culture as well as to individuals who were to be offered for sacrifice.[13] Its "intoxicating" nature made it unpalatable for women and children.[14] Chocolate beverages were also ritually consumed during Mayan betrothal, marriage and baptism ceremonies becoming even more commonly consumed ritualistically in celebratory and commemorative religious gatherings in Aztec culture near its apogee from 1300 to 1521 A.D., especially around Tenochtitlán. Given this use, as W. Tresper Clarke of Rockwood & Co Chocolate Factory in Brooklyn, New York argued, it "seems natural" that the remnants of Chocolate Trees are found along "plazas and streets near ... temples" that were once lined with the "venerable" cacao (Figure 2.4).[15]

Figure 2.3 Indigenous Americans Preparing and Cooking Cacao, as Represented in John Ogilby's *America* (1671), from Brandon Head, *The Food of the Gods* (1903).
(Courtesy of Hershey Community Archives, Hershey, Pennsylvania, USA).

Figure 2.4 Aztecs Carrying Baskets of Tributes to their Gods, from Théodore de Bry's *Histoire de l'Amérique* (1600).

Healers within Mayan cultures belonged to a priestly hierarchy whose members came to their positions via inheritance. Indeed, fine lines distinguished what we would term priests, physicians and sorcerers. As the Mayans were the only pre-Columbian culture that developed writing, their cultural heritage has been recorded in some 750 hieroglyphic signs. The education of Mayan healers, which included the reading and writing of hieroglyphs, also prepared them to practise medicine according to observation, divination and interpretation of a wide variety of omens.[16]

2.2 MESOAMERICAN MEDICAL USES OF CHOCOLATE

The surviving text of a Mayan priest includes chants and incantations offered over sufferers of various diseases including skin disorders, fevers and seizures, during which chocolate mixed with peppers, honey and tobacco juice was also administered.[17] The Badianus Manuscript – a 1552 "Aztec herbal" compiled by the Spanish priests Martín de la Cruz and Juan Badiano – included remedies using cacao flowers to prepare a poultice for sore and injured feet as well as for curing sterility and/or female complaints. Other accounts stress the use of chocolate as an active ingredient in beverage remedies for many stomach disorders, diarrhea, coughs and fatigue. After mixing it with the bark of the silk cotton tree (*Castilla elastica*), it was found to be useful in fighting general infections. Hubert Howe Bancroft, in his *Native Races of the Pacific States* (1875), retold the legend in which Aztecs "dug up the bones of giants at the foot of mountains and collected by their dwarfish successors" would grind them to powder, mix that powder with their cacao, and drink the concoction "as a cure for diarrhea and dysentery".[18]

The Florentine Codex informs us that when an "ordinary amount is drunk, [chocolate] gladdens one, refreshes one, consoles one, invigorates one". From this same record, we also learn that cacao was frequently mixed with a number of other botanicals commonly used in other Mesoamerican medicine preparations in order to enhance the efficacy of the mixture as well as to make the taste of other plant remedies more palatable (Table 2.1).

Bernardino de Sahagún recounts particular associations that the Aztecs held between chocolate and the heart and, relatedly, between chocolate and blood. The terms *yollotl* and *eztli* (heart and

Table 2.1 Botanicals used in Conjunction with Cacao (Florentine Codex).

Ancient Name	*Botanical Name*	Common Name
olquauitl	*Castilla elastica*	Panama Rubber Tree
tlacoxochitl	*Calliandra anomala*	Hummingbird Plant Red Powder Puff
yiauhtli	*Tagetes lucida*	Mexican Tarragon Mexican Mint Marigold
yolloxochitl	*Taluama mexicana Don*	Magnolia
izquixochitl	*Bourreria formosa*	Succulent
teonacaztli	*Cymbopetalum penduliflorum*	Earflower Tree
tecomaxochitl	*Pachira insignis*	Malabar Chestnut
	Calliandra grandiflora	Powder Puff Plant Fairy Duster
	Bombax ellipticum	Shaving Brush Tree
tlilxochitl	*Vanilla planifolia*	Vanilla Orchid

blood, respectively) were synonymous with the Aztec terms for chocolate. The shape of the pods of the Chocolate Tree were often depicted and discussed as being heart-shaped. Both cacao pods and hearts were thought to contain sacred liquids. Select victims who were soon to spill their own blood and perhaps have their living heart vivisected as part of a cultural sacrifice were routinely given *itzpacalatl*, a drink prepared from cacao and blood-stained water, evoking a symbolic connection as well as reputedly sustaining the spirits in these highly agitated individuals.[19] In Mixtec codices, "bleeding cacao pods" were depicted inside and atop temples used for human sacrifice.[20] Annatto, a dye derived from the heart-leaved *achiote* shrub (*Bixa orellana*) was used at the time as a traditional ingredient of chocolate beverages, making them blood red in color. Aztecs commonly mixed cacao with "heart flower" (*Talauma mexicana*) as a remedy for various disorders of the heart.[21] Earlier Mayan culture, as depicted in the Madrid Codex, show hieroglyphs of priests piercing their ears with obsidian lancets, shedding their own blood over cacao pods as a sacrificial ritual to the gods.

Scattered evidence points to the use of cacao among Mesoamerican cultures for a variety of other medical needs. One ancient tradition denotes that cacao helped promote the production of breast milk, thus midwives were usually depicted iconographically carrying cacao beans.[22] The San Blas Indians of Panama also used the pulp in the pods to aid in child delivery. The Seringuerios of

Acre, Brazil used the bark of the Chocolate Tree in concoctions given to women immediately following childbirth.[23] Among its other benefits, Mesoamericans "most often" used cacao as "a soothing agent, antiseptic, stimulant, snakebite remedy, or for weight gain".[24]

Different geographical regions used different parts of *Theobroma cacao* for different aims (Figure 2.5). In Nahau, Mexico, the beans were used to treat people "ill with 'heat'" and to modulate "'hot' body parts such as the liver". These peoples used chocolate in a drinkable form to treat consumption and emaciation. Peoples of the Caribbean used leaves of this plant as a diuretic; those of Venezuela found the leaves helpful in treating burns, eruptions, cracked lips, sore breasts and genitals as well as for vaginal and rectal irritations. The Kuna Indians of Panama and Colombia smoked a preparation of beans, fruit and leaves from this plant to treat malaria and other fevers, whereas the San Blas Indians of Peru burned the beans on the ground, inhaling the smoke as part of the ceremony surrounding the rites of puberty. Many topical uses are found, including easing the extraction of large splinters (French Guinea Indians), a remedy for boils (Quichia of Napo, Ecuador), wound antisepsis (San Blas Indians of Peru), to rid the body of scabies (Ingano), to treat eczema-like lesions of the scalp (Karijona of Colombia) and to make hair grow (Venezuela). The Peruvian Amazons used a drink made of toasted beans to treat dry cough, whereas in Colombia, people used the leaves in an infusion to produce cardiotonic effects and other preparations of salts made from the leaves that likely acted upon the coronary vessels.[25]

Traces of many of Mesoamerican ethnobotanical uses for cacao remain in current use today. For example, an infusion of leaves of *Theobroma cacao* is widely used by the indigenous population of Colombia as a cardiac stimulant and diuretic.[26] A population of Amerinds in Panama use fresh cacao to help stop bleeding.[27] In Central America, *T. cacao* leaves are used to treat screwworm of the eye, whereas different parts of the plant are used in remedies for burns, cough, dry lips, fever, listlessness, rheumatism, snakebite and wound treatment.[28] Reputed antitumor activity of the *T. cacao* root bark has been attributed to the tannin (*i.e.*, polyphenol) content.[29]

Our emphasis on highlighting the medical uses of chocolate in pre-Conquest Mesoamerican cultures should not misconstrue its most common uses. For throughout this vast period, chocolate was

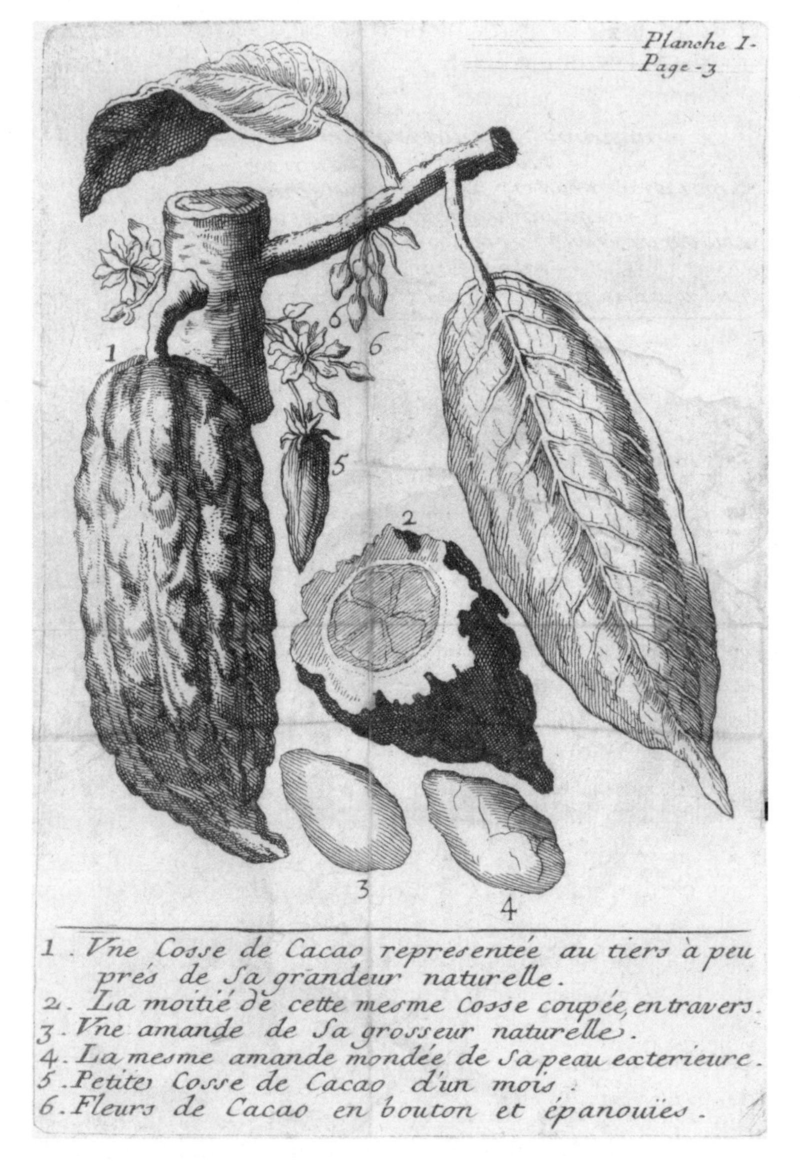

Figure 2.5 *Theobroma cacao*, from Leonhard Ferdinand Meisner's *De Caffe, Chocolatae, Herbae* (1721).
(Courtesy of Hershey Community Archives, Hershey, Pennsylvania, USA).

undoubtedly most often consumed within the context of spiritual and ritualistic experiences, especially as a means of divine supplication.[30]

REFERENCES

1. B. H. Kean, "The blood pressure of the Cuna Indians", *American Journal of Tropical Medicine and Hygiene*, 1944, **24**, pp. 341–343.
2. N. K. Hollenberg, Gregorio Martinez, Marji McCullough, *et al.*, "Aging, acculturation, salt intake, and hypertension in the Kuna of Panama", *Hypertension*, 1997, **29**, pp. 171–176.
3. Further attention to Mesoamerican uses of chocolate are found in Sophie D. Coe and Michael D. Coe, *The True History of Chocolate*, Thames and Hudson, London, 1996, Lisa J. LeCount, "Like water for chocolate: Feasting and political ritual among the late Classic Maya at Xunantunich, Belize", *American Anthropologist*, 2001, **103**, pp. 935–953, Marcy Norton, *Sacred Gifts, A History of Profane Tobacco and Chocolate Pleasures in the Atlantic World*, Cornell University Press, Ithaca and London, 2008, Marcy Norton, "Tasting empire: Chocolate and the European internalization of Mesoamerican aesthetics", *American Historical Review*, 2006, **111**, pp. 660–691, and Cameron L. McNeil, "Introduction: The Biology, Antiquity, and Modern Uses of the Chocolate Tree (*Theobroma cacao L.*)", in *Chocolate in Mesoamerica: A Cultural History of Cacao*, ed. Cameron L. McNeil, University Press of Florida, Gainesville, 2006, pp. 1–30.
4. *Cacao* to the Maya was originally phonetically pronounced by the Olmec's as *kakawa*.
5. Teresa Dillinger, Patricia Barriga, Sylvia Escárcega, *et al.*, "Food of the Gods: Cure for humanity? A cultural history of the medicinal and ritual use of chocolate", *JN The Journal of Nutrition*, 2000, **130 Supplement**, pp. 2057S–2072S, p. 2058S.
6. These chronicles comprise what is now called the Florentine Codex. Quetzalcoatl reputedly acquired cacao from The Garden of Life. Catholic missionaries deemed the content of codices as perpetuating heathen traditions, and many of these historical records were destroyed. For instance, Diego de Landa Calderón, Spanish Bishop of the Roman Catholic Archdiocese of Yucatán, is thought to have destroyed at least 27 codices in 1562.
7. Pierre Labane, "The History of Chocolate", in *The Book of Chocolate*, Nathalie Bailleux, Hervé Bizeul, John Feltwell,

et al., Flammarion, Paris and New York, 1996, pp. 59–104, p. 63.

8. Carole Bloom, *All About Chocolate: the Ultimate Resource to the World's Favorite Food*, Macmillan, New York, 1998, p. 164.
9. Nisao Ogata, Arturo Gómez-Pompa and Karl A. Taube, "The Domestication and Distribution of *Theobroma cacao L.* in the Neotropics", in *Chocolate in Mesoamerica: A Cultural History of Cacao*, ed. Cameron L. McNeil, University Press of Florida, Gainesville, 2006, pp. 69–89 have argued that the pod structures depicted on a 2500-year-old Peruvian vessel most likely represent cacao.
10. W. Jeffrey Hurst, Robert A. Martin, Jr., Stanley M. Tarka, Jr. and Grant D. Hall, "Authentication of cocoa in Maya vessels using high-performance liquid chromatographic techniques", *Journal of Chromatography*, 1989, **466**, pp. 279–289; Grant D. Hall, Stanley M. Tarka, Jr., W. Jeffrey Hurst, *et al.*, "Cacao residues in ancient Maya vessels from Rio Azul, Guatemala", *American Antiquity*, 1990, **55**, pp. 138–143; Cameron L. McNeil, "Introduction: The Biology, Antiquity, and Modern Uses of the Chocolate Tree (*Theobroma cacao L.*)", in *Chocolate in Mesoamerica: A Cultural History of Cacao*, ed. Cameron L. McNeil, University Press of Florida, Gainesville, 2006, pp. 1–30; and W. Jeffrey Hurst, "The Determination of Cacao in Samples of Archaeological Interest", in *Chocolate in Mesoamerica: A Cultural History of Cacao*, ed. Cameron L. McNeil, University Press of Florida, Gainesville, 2006, pp. 105–113.
11. Grant D. Hall, Stanley M. Tarka, Jr., W. Jeffrey Hurst, *et al.*, "Cacao residues in ancient Maya vessels from Rio Azul, Guatemala", *American Antiquity*, 1990, **55**, pp. 138–143, p. 141.
12. W. Jeffrey Hurst, Stanley M. Tarka, Jr., Terry G. Powis, *et al.*, "Archaeology: Cacao usage by the earliest Maya civilization", *Nature*, 2002, **418**, pp. 289–290. For anthropological insights into the preparation of chocolate in terms of social relations and social stratification, see Lisa J. LeCount, "Like water for chocolate: Feasting and political ritual among the late Classic Maya at Xunantunich, Belize", *American Anthropologist*, 2001, **103**, pp. 935–953 and Rosemary A. Joyce and John

S. Henderson, "From feasting to cuisine: Implications of archaeological research in an early Honduran village", *American Anthropologist*, 2007, **109**, pp. 642–653.

13. As in so much history, there are no surviving records (if they were ever kept) of food and ritual among the common class of people. Still, the mere lack of this evidence does not absolutely translate to a "fact" that chocolate was not consumed, at least ritualistically, by the common classes of the social hierarchy.
14. Teresa Dillinger, Patricia Barriga, Sylvia Escárcega, *et al.*, "Food of the Gods: Cure for humanity? A cultural history of the medicinal and ritual use of chocolate", *JN The Journal of Nutrition*, 2000, **130 Supplement**, pp. 2057S–2072S, p. 2058S. As Betty Bernice Faust discussed, an exception to this may have been the use of cacao beans symbolically and physiologically in ceremonies surrounding the onset of menarche in Mayan youth.
15. W. Tresper Clarke, "The Literature of Cacao", in *Advances in Chemistry*, American Chemical Society, 1954, **10**, pp. 286–296, p. 287.
16. For an overview of the Maya medical heritage use of both natural and supernatural remedies, see Francisco Guerra, "Maya medicine", *Medical History*, 1964, **8**, pp. 31–43 as well as Marianna Appel Kunow, *Maya Medicine: Traditional Healing in Yucatán*, University of New Mexico Press, Albuquerque, 2003 that also incorporates anthropological investigations of ancient medicine based, in part, upon the long-standing healing traditions used today. Francisco Guerra, "Aztec medicine", *Medical History*, 1966, **10**, pp. 315–388 also provides an overview of Aztec medicine.
17. Ruth Lopez, *Chocolate: The Nature of Indulgence*, H. N. Abrams in association with the Field Museum, New York, 2002, p. 64. The text, known as the *Ritual of Bacabs*, was discovered in the Yucatán and is now at Princeton University. The Bacabs were four divine brothers placed at the four corners of the world where they held up the skies. This text also includes many "magical" components of Maya medicine.
18. Chantal Coady, *Chocolate: The Food of the Gods*, Chronicle Books, San Francisco, 1993, p. 76.

19. Sarah Moss and Alexander Badenoch, *Chocolate: A Global History*, Reaktion Books, London, 2009, pp. 17–18.
20. Cameron L. McNeil, "Introduction: The Biology, Antiquity, and Modern Uses of the Chocolate Tree (*Theobroma cacao L.*)", in *Chocolate in Mesoamerica: A Cultural History of Cacao*, ed. Cameron L. McNeil, University Press of Florida, Gainesville, 2006, pp. 1–30, p. 15.
21. Meredith L. Dreiss and Sharon Edgar Greenhill, *Chocolate: Pathway to the Gods*, University of Arizona Press, Tucson, 2008, pp. 146–147. Curiously, as these authors remind us, this is still used as a folk remedy throughout rural Mexico among cardiac patients to produce a digitalis (Foxglove)-like effect, increasing cardiac contractility and better controlling heart rate.
22. Cameron L. McNeil, ed., *Chocolate in Mesoamerica: A Cultural History of Cacao*, University Press of Florida, Gainesville, 2006, pp. 364–365.
23. Nathaniel Bletter and Douglas C. Daly, "Cacao and Its Relatives in South America", in *Chocolate in Mesoamerica: A Cultural History of Cacao*, ed. Cameron L. McNeil, University Press of Florida, Gainesville, 2006, pp. 31–68, pp. 53, 55.
24. Nathaniel Bletter and Douglas C. Daly, "Cacao and Its Relatives in South America", in *Chocolate in Mesoamerica: A Cultural History of Cacao*, ed. Cameron L. McNeil, University Press of Florida, Gainesville, 2006, pp. 31–68, p. 48.
25. Nathaniel Bletter and Douglas C. Daly, "Cacao and Its Relatives in South America", in *Chocolate in Mesoamerica: A Cultural History of Cacao*, ed. Cameron L. McNeil, University Press of Florida, Gainesville, 2006, pp. 31–68, pp. 52–55.
26. R. Shultes and R. Raffauf, *The Healing Forest*, Dioscorides Press, Portland, 1990, p. 447.
27. M. P. Gupta, P. N. Solís, A. I. Calderón, *et al.*, "Medical ethnobotany of the Teribes of Bocas del Toro, Panama", *Journal of Ethnopharmacology*, 2005, **96**, 389–401.
28. James A. Duke, *Isthmian Ethnobotanical Dictionary*, 3rd edn, Scientific Publishers, Jodphur, India, 1986, p. 188, and J. A. Duke, *CRC Handbook of Medicinal Herbs*, CRC Press, Boca Raton, FL, 1985, p. 479.
29. J. A. Duke, *CRC Handbook of Medicinal Herbs*, CRC Press, Boca Raton, FL, 1985, p. 479.

30. For insight into specific changes in the ritualistic use of chocolate in Santiago de Guatemala (present-day Antigua), the capital city of colonial Central America during slightly later periods, see Martha Few, "Chocolate, sex and disorderly women in late-seventeenth- and early-eighteenth-century Guatemala", *Ethnohistory*, 2005, **52**, pp. 674–687. See also Betty Bernice Faust, "Cacao beans and chili peppers: Gender socialization in the cosmology of a Yucatec Maya curing ceremony", *Sex Roles*, 1998, **39**, pp. 603–641. As for chocolate in New Spain more from a cookery perspective, see Rachel Laudan and Jeffrey M. Pilcher, "Chiles, chocolate, and race in new Spain: Glancing backward to Spain or looking forward to Mexico?", *Eighteenth-Century Life*, 1999, **23**, pp. 59–70, as well as a broader review of consumption by Marcy Norton, *Sacred Gifts, A History of Profane Tobacco and Chocolate Pleasures in the Atlantic World*, Cornell University Press, Ithaca and London, 2008.

CHAPTER 3

Cacao Transported to Europe as Medicine

> As chocolate grew in popularity, it gained a reputation for having certain powers. Physicians began prescribing chocolate as a bromide, a cure-all, for their patients. Others, less concerned with its curative abilities, suggested chocolate's primary virtue to be that of a stimulant, and even more particularly, an aphrodisiac. Whatever the goal happened to be, chocolate was the way to get there.
>
> Ray Broekel, *The Chocolate Chronicles* (1985)[1]

Chocolate first became widely documented as a medicinal beverage in the Old World during the 1600s. Just when the bean made its trans-Atlantic migration and became known to various groups of Europeans remains shrouded in mystery. During the Genoese Christopher Columbus' 4th and final voyage to the Indies (1502–1504), his son Ferdinand reported their encounter on 15 August 1502 with a hoard of "almonds of money" stored in the huge Mayan trading canoe in the gulf of Honduras which Columbus and his men had captured.[2] Natives treasured this money such that when an "almond" (*i.e.*, cacao bean) fell to the ground, they "all stooped to pick it up, as if an eye had fallen".[3]

No evidence survives as to whether Columbus and his voyagers ever drank a concoction prepared from this treasured bean.

Chocolate as Medicine: A Quest over the Centuries
Philip K. Wilson and W. Jeffrey Hurst

Published by the Royal Society of Chemistry, www.rsc.org

Despite widely rumored accounts, no evidence supports the claims that Spanish *conquistador* and invader Hernán Cortés, who established a cacao plantation in Mexico, widely introduced cacao ("brown gold") to Europe on his 1528 return from the New World.[4] Still, Cortés is rightly credited with bringing samples of this substance to King Charles of Spain and first reporting that, in the Americas, cacao was used to prepare a "divine drink" that "builds resistance and fights fatigue".[5]

Knowledge of this bean's use in preparing chocolate drinks appears to have spread more widely following 1544 when Dominican friars of the Aragón Monastery (including the scholarly Bartolomé de Las Casas, confidant to Spanish King Charles' son, Philip), who had befriended a Guatemalan delegation of Kekchi Mayan nobles in Spain, shared stories of a drink that they had been enjoying within their cloistered monasteries.[6] Curiously, this spiritually minded group consumed cacao merely as a tasty and nutritious drink and drug without considering it to hold any spiritual powers as it had for the Mesoamericans. Though a gustatory bond had formed between the New World and the Old, chocolate's medical benefits had shifted from a focus upon holistic, spiritual health to that of physical, secular health during this trans-Atlantic migration. Still, intrigue over the somewhat mystical or magical power that chocolate just might contain remained part of its persistent allure.

In this era, news and new products travelled slowly. It was not until near the century's end, in 1585, that the first specifically designated shipment of cacao reached Europe, travelling from modern day Veracruz to Seville (Figure 3.1). Only three years prior to this, the image of a cacao bean and a description of the Chocolate Tree first appeared in print in Europe, that being the botanist and horticulturalist Carolus Clusius's account of the plants gathered in Sir Francis Drake's recent circumnavigation of the globe.[7] As food historian Reay Tannahill argued in *Food in History* (1973), for "well over a hundred years" since *conquistadores* first learned of the chocolate drink, cacao beans – a "mirror of trade and conquest" – became Spain and Portugal's "jealously guarded monopoly".[8] Some believe that the eventual disruption of this veritable monopoly "coincided with the decline of Spain as a major world power".[9]

Figure 3.1 Neptune receiving Chocolate from The Americas to Transport to Europe. Frontispiece of Antonio Colmenero de Ledesma, *Curioso Tratado de la Naturaleza y Calidad del Chocolate* (1631), from "Historicus", *Cocoa: All About It* (1892).

Touting this treasured botanical find, the Jesuit priest, Alonsius Ferronius composed the following "Ode to The Chocolate Tree" in 1664 and dedicated it to Cardinal Francesco Maria Brancaccio.

O tree, born in far off lands,
Prize of Mexico's shores,
Rich with a heavenly nectar
That will conquer all who taste it.

To thee let every tree pay homage,
And every flower bow its head in praise
The wreath of the laurel crowns you; the oak, the alder,
And the precious cedar proclaim your triumph.

Some say you lived in Eden with Adam
And that he carried you with him when he fled.
And from thence you journeyed to the Indies
Where you prospered in the hospitable soil,
And your trunk burgeoned with
The bounty of your noble seeds.

Are you another gift from Bacchus,
Famed for his free-flowing wines?
No – the fruits of Crete and Massica
Bring not the glory you do to your native land.

For you are a fresh shower that bedews the heart,
The fountain of a poet's gentle spirit.
O sweet liquor sent from the stars.
Surely you must be the drink of gods![10]

3.1 PREPARING CHOCOLATE REMEDIES

For decades, rumors abounded as to just how the Mesoamericans prepared their healthy concoction from these bitter tasting beans. The Florentine circumnavigator, Francesco d'Antonio Carletti reported from his late 16th century travels that in Mexico, bricks or blocks made from cacao were turned into a paste that was mixed with spices and water and then frothed. A much later account notes that the name "chocolate" referred to the process of turning cacao into a drink. As an "Indian word", it is "compounded of *Choco*" meaning sonus or sound, and *Atte* or *Atle* meaning water (*aqua*), so named "because they commonly make use of Water to prepare Chocolate ... and make a little rusling [sic] with an Instrument called a Chocolate-stick [*molinet* or *molinillo*], which is made use of to stir it" (Figure 3.2).[11] Carletti continues that once prepared, this mixture is gulped

> in one swallow with admirable pleasure and satisfaction of the bodily nature, to which it gives strength, nourishment and vigor in such a way that those who are accustomed to drinking it cannot remain robust without it even if they eat other substantial things. [Furthermore, these individuals ...] appear to diminish when they do not have that drink.[12]

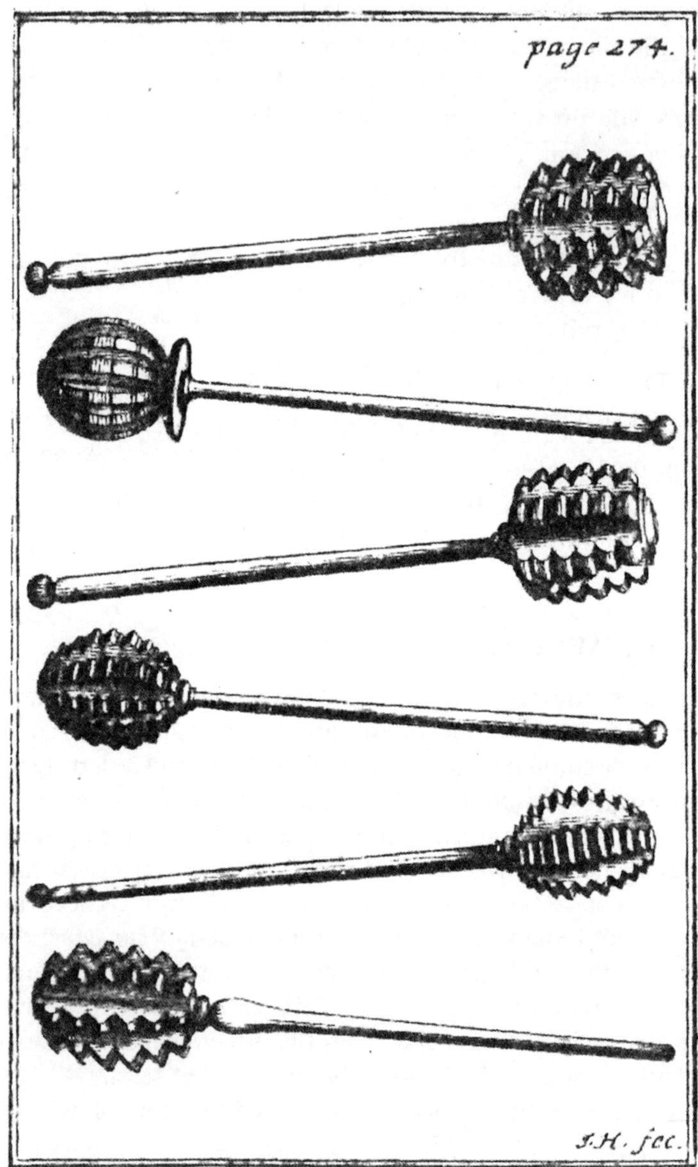

Figure 3.2 Diverse Forms of *Molinets*, from Nicholas de Blégny's *Le Bon Usage du Thé, du Caffé et du Chocolat* (1687).

The Jesuit priest José de Acosta in his *Historia Natural y Moral de las Indias* (1590) noted a similar dependency upon cacao, writing that "Spanish men – and even more the Spanish women – are addicted" to chocolate.[13] Dependency, perhaps, but as the Milanese traveller and historian Girolamo Benzoni noted in his *La Historia del Mondo Nuovo* (1565), unlike wine, the chocolate drink "satisfies and refreshes the body, but does not inebriate".[14]

Not all chocolate drinks were the same, however. Juan de Cárdenas reported in 1591 that vast differences resulted from drinks prepared with "green" (*i.e.*, untoasted) beans versus toasted beans. Drinks prepared with untoasted beans could harm digestion by obstructing the liver, spleen and bowels, creating paroxysms, melancholy and irregular heartbeats, shortening the breath and depriving the face of its life and natural color, whereas those with toasted beans (a process thought to increase the oily constituency of the beans) aided digestion, built strength and became fattening and sustaining to those who partook of this medicinal beverage.[15] Botanist and "First Chief Physician of all Indes, Islands, and Lands of the Sea Ocean", Francisco Hernández de Toledo reported to Philip II of Spain his New World observations during the 1570s that natives often mixed concoctions containing equal portions of cacao beans and seed of the ceiba tree, the special tree of Mayan culture that linked the terrestrial and spiritual worlds through its long hanging vines whereby souls left the earthly realm.[16] Cane sugar, when available, was used in the New World to sweeten the drink to taste. The sweetened chocolate drink quickly became the preferred taste, after which the cost of preparing this healthy beverage typically fluctuated with the price of sugar.

Perhaps the greatest difference between New and Old World habits for sipping chocolate was the preferred temperature of their drink. The New World chocolate was a cold drink, whereas it was modified in Spain as a hot drink. Though this may seem a trivial distinction, this preference reflected distinct views of bodily constitutions and temperaments that had remained at the core of Old World medical wisdom since the ancient and influential writings comprising the Hippocratic *Corpus* as well as in the works of Galen, the most renowned medical philosopher and physician-writer of ancient Rome.

For these medical authorities, disease generally owed its origin to natural causes. In particular, diseases occurred from an imbalance of the natural order or constitutional makeup (*i.e.*, temperament) of the body. The specific imbalance was thought to have been caused by particular rearrangements of the four bodily humours. The ancients viewed the universe as composed of four elements: air, earth, fire and water. These four elements were linked to four humours in the body, respectively: blood, black bile, phlegm and yellow bile. The human body was considered to be in a state of disease, more specifically of dis-ease, when the equilibrium that naturally existed between the humours was imbalanced. To further account for states of dis-ease as well as to prepare appropriate treatment regimens, each humour was also described as being comprised of two qualities, wet or dry and hot or cold.

It is in relation to these humoural qualities that many early discussions of chocolate's medicinal benefits were formulated. For example, Hernández argued that because the cacao bean was itself viewed to hold cold and dry qualities, drinks prepared from it were best designated as a curative for disorders of opposite humoural qualities. That is, cacao beverages were most commonly used to treat fevers and other diseases identifiable by their hot and wet qualities.[17] Relatedly, chocolate was also a drink found to benefit those living in hot climes. Antonio Colmenero de Ledesma warned that as cacao was cold and dry, it was apt to produce melancholy (a humoural state caused by an excessive concentration of black bile) on its own. Dr Juanes de Barrios amended this view, explaining that by mixing it into special concoctions, chocolate could actually help people of all temperaments. Melancholics (*i.e.*, those of cold and dry, black bile predominant temperaments) were told to sip a lukewarm chocolate drink prepared without chilies but with a few anise seeds. Phlegmatic (cold, wet, phlegm predominant) temperament patients were given a hot and spicy chocolate drink. Sanguine (warm, wet, blood predominant) temperament patients should be given chocolate drink prepared without corn flour, whereas those choleric (warm, dry, yellow bile predominant) patients should receive "milder, more temperate" forms of chocolate drink.

Overheated people were also thought to benefit from chocolate, though it was recommended that mixtures used to treat their conditions be prepared with spices of a hot nature including cinnamon, anise seed and black pepper that tended to balance the otherwise cold nature of cacao. Such hot-spiced chocolate drinks were additionally noted to help "warm the stomach, perfume the breath, combat poisons, alleviate intestinal pain" as well as to "excite venereal passion".[18] In 1624, Joannes Franciscus Rauch also noted this last consequence, and he framed his entire *Disputatio Medico Dioetetica de Aëre et Esculentis, de Necnon Potu* as a condemnation against cacao, a substance he found to be a "violent inflamer of passions".[19] The English physician Henry Stubbe attributed this "provocative to lust" to a view that chocolate "begets good blood".[20] Decades before Richard Cadbury first put chocolate in a heart-shaped box for Valentine's Day in 1861, chocolate was being widely promoted by some (and likely prohibited by others) for its libido-enhancing effects.[21]

Various modulations of cacao preparations were deemed necessary in the Old World since people from different geographical regions were believed to be composed of different temperaments. New World Indians, for example, were thought to be able to "stomach" cacao mixed with water quite well due to having been raised in the heat-filled country, whereas similar concoctions for Old World chocolate drinkers frequently produced stomach aches and impeded digestion. English physician and natural philosopher Martin Lister, upon visiting Paris in 1698, expressed a similar concern over chocolate, noting that "Europeans, unlike the Indians, do not have the ability to digest it, and that drinking chocolate leads to a worn-out, decaying gut".[22]

3.2 CHOCOLATE IN THE MEDICAL LITERATURE

In effort to popularise cacao in the New World, a number of monographs appeared throughout the 1600s that described this still new and somewhat mysterious substance. Antonio Colmenero de Ledesma's *Curioso Tratado de la Naturaleza y Calidad del Chocolate* (1631) – the first book devoted entirely to chocolate – spread claims of this complex and healing substance more widely through its translations into English, French, Latin and Italian. The title page of this work, in its English translation, attests to chocolate's perceived medicinal benefits: by the "wise and moderate use whereof health is preserved, sicknesse diverted, and cured, especially the plague of the guts; vulgarly called the new disease;

fluxes, consumptions, and coughs of the lungs, with sundry other desperate diseases. By it also, conception is caused, the birth hastened and facilitated, beauty gain'd and continued". Fortunate for readers, this translation by Captain James Wadsworth was available in London near the Vine Tavern in Holborne as was "the chocolate itself" – and according to a contemporary advertisement, both "may be had at reasonable rates".[23]

Wadsworth, under the pseudonym Don Diego de Vadesforte, added a poem to the preface of this work in which he stressed the seemingly universal health benefits of chocolate. It read, in part (full text in Appendix 4), as follows:

Doctors lay by your Irksome Books
And all ye Petty-Fogging Rookes
Leave Quacking; and Enucleate
The vertues of our Chocolate.

Let th' Universall Medicine
(Made up of Dead-mens Bone and Skin,)
Be henceforth Illegitimate,
And yield to Soveraigne-Chocolate....

Tell us no more of Weapon-Salve,
But rather Doome us to a Grave;
For sure our wounds will Ulcerate,
Unlesse they're wash'd with Chocolate....

The Roaring-Crew of Gallant-Ones
Whose Marrow Rotts within their Bones;
Their Bodyes quickly Regulate,
If once but Sous'd in Chocolate....

Nor need the Women longer grieve
Who spend their Oyle, yet not conceive,
For 'tis a Helpe-Immediate,
If such but Lick of Chocolate....

The Feeble-Man, whom Nature Tyes
To doe his Mistresse's Drudgeries;
O how it will his mind Elate,
If shee allow him Chocolate![24]

Among England's chief "chocoholics" was the physician, Henry Stubbe. His 1662 book, *The Indian Nectar; or, A Discourse Concerning Chocolata*, was specifically designed to help the English reading public overcome some common misconceptions regarding the strength and frequency of using chocolate as a medicine. To support his claims, Stubbe relied upon case histories drawn from the lands of cacao's origin as the most solid form of evidence. "English soldiers stationed in ... Jamaica lived [for many months on only] cacao nut paste mixed with sugar ... which they [drank having] dissolved [it] in water". Women of the New World were also reported as having eaten chocolate "so much ... that they scarcely consumed any solid meat yet did not exhibit a decline in strength".[25]

Stubbe provided additional evidence from case studies of reputable New World physicians. For example, he recorded the indigenous medical wisdom that chocolate is

> one of the most wholesome and pretious [sic] drinks, that [has] been discovered to this day: because in the whole drink there is not one ingredient put in, which is either hurtful in it self, or by commixtion; but all are cordial, and very beneficial to our bodies, whether we are old, or young, great with child, or ... accustomed to a sedentary life.[26]

Citing the work of Dr Juanes de Barrios, Stubbe noted the seemingly common belief that chocolate was "all that was necessary for breakfast, because after eating chocolate, one needed no further meat, bread or drink".[27] Given the great healthiness resulting from chocolate consumption, Stubbe described the precise blends of chocolate mixed with other agents that were found to be most helpful in treating particular ailments.

Further information regarding New World uses of chocolate is found in William Hughes' *The American Physitian [sic], or a Treatise of the Roots, Plants, Trees, Shrubs, Fruit, Herbs &c. Growing in the English Plantations in America; with a Discourse on the Cacao-Nut-Tree ... and All the Ways of Making of Chocolate* (1672).[28] Hughes penned his narrative after serving aboard a ship to the West Indies where he became well acquainted with American herbs and their medical uses. In it, he noted that the inhabitants of Nicaragua, New Spain, Mexico, Cuba and Jamaica so highly treasured the powers derived from the pods of the Chocolate Tree, that they took extreme measures to secure these plants within the "shades of Plantane [sic] and Bonona [sic] Trees, against the injuries of their fiery Sun ... " (Figure 3.3). He also noted anecdotally that Moctezuma is said to have treated Cortés and his soldiers with it, and that one can "scarce read an American Traveller" account that does not "tell you of the magnificent Collations of Chocolate, that the Indians offer'd him in his Passage and Journies [sic] through their Country".[29]

As a specific example of chocolate's medicinal properties, Hughes described how

> Indians and Christians, in the American Plantations, have been Observ'd to live several Months upon Cocoa [sic] Nuts alone, made into a Paste with Sugar, and so dissolv'd in Water;

Figure 3.3 Shading the Chocolate Tree, the Earliest Printed Engraving of Cacao which Appeared in Giralamo Benzoni's *La Historia del Mondo* (1565), from C.J.J. Van Hall's *Cacao* (1914).

> I myself have eaten great quantities of these Kernels raw, without the least inconvenience; and have heard that Mr [Robert] Boyle and Dr [Henry] Stubbe, have let down into their Stomachs some Pounds of them raw without any molestation; the Stomach seems rather to be satisfied than cloy'd with them, which is an Argument they are soon dissolv'd and digested.[30]

Hughes informed his readers, drawing largely upon anecdotal information as well as upon what he termed "Experimental Observations", that "curious Travellers and Physicians do agree" that chocolate "has a wonderful faculty of quenching thirst, allaying Hectick Heats, [and] of nourishing and fatning [sic] the Body".[31] He related how the English Dominican Friar Thomas Gage who, following his education in Xeres, Spain, after having spent time in the New World with Hernán Cortés, "acquaints us, that he drank Chocolate in the Indies [Chiapas, New Spain] two or three Times every Day, for twelve years together, and he scarce knew what any Disease was in all that time"; the only noticeable effect was that he grew "very fat".[32] Others expressed disdain over the use of pure chocolate, which they considered as "too oily and gross". Still, they admitted that "the bitterness of the Nut makes amends, carrying the other off by strengthening of the Bowels" (Figure 3.4).[33]

Hughes offered his own personal narrative as evidence, informing readers that "he liv'd, at Sea for some Months" on "nothing but Chocolate, yet neither his strength, nor flesh were diminished".[34] Indeed, like Friar Gage, he "grew very fat in Jamaica by vertue of the Cocoa [sic] Nut".[35] Accordingly, he claimed it to be of considerable help in counteracting "Lean, Weak, and Consumptive Complexions". He also noted that it "may be proper for some breeding Women, and those persons that are Hypochondriacal, and Melancholly [sic]".[36] Elaborating further, Hughes noted chocolate, when "internally administered" to be "good against all coughs, shortness of breath, opening and making the roughness of the Artery smooth", thereby "palliating all sharp Rheums, and contributing very much to the Radical Moisture, being very nourishing, and excellent against Consumptions". The "fat Butter or Oyl" of the cacao bean was reputed to be "very effectual, being externally applied, against all inflammations, *i.e.*, Phlegmons,

Figure 3.4 Friar Thomas Gage Receiving his Parishioners' Offerings, from Gage's *A New Survey of the West-Indies* (1648).

Erysipelas, St Anthony's Fire, Smallpox, Tumours, Scaldings and Burnings". Externally, it was found to cool the "pains proceeding from heat" as in the "crustiness or scars on Sores, Pimples, chapped Lips and Hands" as well as working to "wonderfully refresheth wearied limbs" and to "mitigate the pain of the Gout, and also Aches by reason of old Age".[37] To rhetorically boost his own narrative, Hughes added that medicines "whereof the cacao is the principal ingredient" had become "approved of by Learned Physicians, and sufficiently recommended to the world".[38]

Hughes' contributed greatly to the spread of the American indigenous use of plants "either for Meat or Medicine". In this work, he elaborated more upon cacao – the "American Nectar" – than upon any other plant. Such contributions were commemorated poetically in "On Mr Hughes's Treatise of American Plants", a work attached to later editions of his work:

> The world of Treatises hath had great store,
> But such an one was never seen before:
> What here's disclos'd, Columbus did not see
> In his American Discoverie.
> He to find out the Land did boldly venture;
> But Hughes i'th'bowels of the Land did enter,
> To finde the Roots of Plants, and rarer things,
> To profit Subjects, and to please their Kings.
> Our Lovel, Gerrard, Johnson, and learn'd Ray
> Did travel far in the Botanick way:
> But this our Author hath out-went them clear,
> As by the following lines it doth appear:
> In which the Plants of India may be found,
> And their Vertues, to keep our Bodies sound.[39]

Throughout Europe, case study evidence became an increasingly used form of rhetoric during the Early Modern Period to convince physicians and the public of chocolate's perceived benefits. Philippe Sylvestre Dufour (perhaps a pseudonym of Jacob Spon) provided such evidence in his popular and widely translated, *De l'Usage du Caphé, du Thé, et du Chocolat* (1671).[40] Dufour also noted one of the commonly reported side effects of considerable chocolate drinking: people got a bit larger over time.

> The "buttery parts" of the cacao tend to "fatten" people because "the 'hot ingredients' of medicinal chocolate serve as a type of pipe or conduit ... and make it pass by the liver, and the other parts till they arrive at the fleshy parts, where finding

a substance which is like and comfortable to them, ... [they] convert themselves into the substance of the subject [whereby] they augment and fatten it".[41]

A steady stream of writings on chocolate and health began to roll off publishers' presses. Henry Mundy prepared a work on chocolate particularly for medical audiences, *Opéra Omnia Medico-Physica de Aëre Vitali, Esculentis et Potulentis cum Appendice de Parergis in Victu et Chocolatu, Thea, Caffea, Tobacco* (1685) as did Marcus Mappus with his *Dissertationes Medicae Tres de Receptis Hodie Etiam in Europa, Potus Calidi Generibus Thée, Café, Chocolata* (1695). Royal physician, Nicolas de Blégny offered case study evidence supporting chocolate's use in maintaining and restoring soldiers' health in his *Le Bon Usage du Thé, du Caffé, et du Chocolat pour la Préservation et pour la Guérison des Maladies* (1687).

Chocolate's spread throughout Europe was notably based upon its use as a medication. Italian physician Paolo Zacchia, in his 1644 writings on the hypochondriac diseases, specifically referred to chocolate as a drug. The Cardinal of Lyon, Alphonse-Louis du Plessis de Richelieu (brother of the famed Armand Jean du Plessis, Cardinal-Duc de Richelieu et de Fronsac), introduced chocolate to France in 1651 as a drug to "moderate the vapours of the spleen" – actions that he attested from his personal use.[42] He also noted that chocolate was useful in helping people overcome bouts of anger and fits of bad temper.

3.3 PATIENT ACCOUNTS

Occasionally, very occasionally, cases were recorded from the view of the patient. Madame Marie de Villars, wife of the French Ambassador to Spain claimed that it was her steady consumption of chocolate that brought her great health. In the 1660s, Maria Theresa (daughter of Spain's King Philip IV and wife of French King Louis XIV) prompted a nationwide passion for chocolate; her physician, Dr Joseph Bachot, promulgated it as the "true food of the gods".[43] In England, following the festivities celebrating Charles II's Coronation, the diarist Samuel Pepys noted, "Waked in the morning with my head in a sad taking through the last night's drink, which I am very sorry for; so rose, and went out with Mr Creed to drink our morning draught, which he did give me in chocolate [what Pepys commonly referred to as *jocalette*] to settle my stomach".[44] A Dutchman visiting French Martinique in 1720 was regaled with the story of a councilor "of about a hundred years of age" who had recently died. According to a story that this visitor later shared, the Councilor had "subsisted for thirty years on nothing other than chocolate and some biscuits. Occasionally, he would take a little soup to eat, but at no time meat, fish or other nourishment. Yet, he was so fit that, at the age of eighty-five years, he could still mount his horse without stirrups".[45]

A series of letters from 1671 provides a glimpse into another patient, Marie de Rabutin-Chantal, Marquise de Sévigné's changing views of chocolate's therapeutic potential, giving insight into a range of fashionable fads in chocolate's use over time. In February, writing to her daughter, Françoise-Marguerite de Sévigné, Comtesse de Grignan, who was pregnant at the time, she noted a strong favorable view, "But you are not well, You have hardly slept, Chocolate will set you up again". Two months later, her enthusiasm for chocolate had considerably diminished, claiming it to be "no longer for me what it was... Everyone who spoke well of it now tells me bad things about it ... causing one's ills, it is the source of all vapors and palpitations ... it suddenly lights as a continuous fever in you that leads to death". But, as with the change of seasons, she regained her support, noting, in October, that chocolate "acts cording [sic] to my intentions".[46] Like many of the earliest New World geographical and travel narratives, the patient narratives of chocolate as medicine are typically based upon primarily anecdotal evidence.

Reaching back to the earliest recordings of chocolate, accounts of its use also typically describe at least rudiments of chocolate preparations. The Florentine Codex details the elaborate ritualistic preparations required to release cacao's therapeutic potential. Early travel narratives, such as Francisco Hernández's *Historia de las Plantas de la Nueva España* (1577), described how cacao's benefits were enhanced by mixing it with specific other ingredients locally obtainable in particular regions. Stubbe, in *The Indian Nectar*, argued that all of the ingredients added to chocolate remedies must be precisely correlated with distinct individual constitutions in mind. In addition to cacao itself, he noted,

> the other Ingredients for making up Chocolate ... [must] be varied according to the Constitutions of those that are to

> drink it; in cold Constitutions, Jamaica Pepper, Cinnamon, Nutmegs, Cloves, *&c.* may be mixt with the Cacao nut; some add Musk, Ambergrease,[47] Citron, Lemon-peels, and Odoriferous Aromatick Oyls. In hot Consumptive tempers you may mix Almonds, Pistachos, ... sometimes China, Sarfa, and Saunders; and sometimes Steel and Rhuburb may be added for young green Ladies.

In sum, Stubbe concluded,

> [So] that you may know how to Prepare your Chocolate, I will give you a short direction, – if you intend to make it up yourself, consult your own Constitution and Circumstances, and vary the Ingredients according to the Premises.[48]

3.4 CHOCOLATE IN THE PHARMACY LITERATURE

In Early Modern Europe, an entire growing body of newly assembled sources began to shift the focus of chocolate as medicine more towards an in-depth analysis of specific chocolate concoctions than upon either mere anecdotal or case-study evidence. New herbals, pharmacopoeias, formularies and other works of *materia medica* drew upon many regions and cultures in their formulaic compilations of remedies. In such works, typically designed for professional use, the specific ingredients required for newly designed chocolate compounds were referenced alongside those of more long-standing use. One of the most widely translated of these new types of medical writing was Nicolas Lémery's *Traité Universel des Drogues Simples* (1698). In it, we learn that,

> Chocolate is nourishing enough. It is strengthening, restorative, and apt to repair decay'd Strength, and make People strong. It helps Digestion, allays the sharp Humours that fall upon the Lungs. It keeps down the Fumes of Wine, promotes Venery, and resists the Malignity of the Humours.
>
> When Chocolate is taken to Excess, or that you use a great many sharp and pungent Drugs in the making of it, it heats much, and hinders several People to sleep.

Guided by the medical wisdom of the four bodily humours, readers who were also those most likely to compound chocolate concoctions came to appreciate that,

> Chocolate agrees, especially in cold Weather, with old People, with cold and phlegmatic Persons, and those that cannot easily digest their Food, because of the Weakness and Nicety of their Stomachs; but young People of a hot and bilious Constitution, whose Humours are already too much in Motion, ought to abstain from it, or use it very moderately.[49]

Acknowledgement is given to "the Americans" who "shew'd the Way of making it to the Christians", but readers of this French treatise were strongly encouraged to consider that "the Chocolate made at Paris" was greatly "improved ... by the Compositions we use" (Figure 3.5). The specific ingredients and formulaic steps of compounding remedies were offered, whereupon a product would be created that could be "eaten as is" or drank "after dissolving it in ... Common Water ... [or] Cows Milk, ... [or] Almond-Milk" or in the "juice ... of Plants" or mixed with "a little Bezoar stone ... to make it more Cordial".[50]

Figure 3.5 French Artisanal Chocolate Shop Operations, *Encyclopédie* (1715). Dried Cacao Beans are being 1) Roasted, 2) Shelled (Winnowed), 3) Ground, and 4) Rolled into Paste.

Lémery summarised chocolate's medicinal benefits, claiming that it would

> [H]elp Digestion, recover decay'd Strength, and produce a great many the like Effects. It may be also good for phthisical [tubercular] People ... [particularly] because the Cacao-nut ... being full of oily and balsamic Principles, is ... very good for allaying and embarrassing the sharp Humours, which are predominant in those that are troubled with the Phthisick [sic], and for nourishing and recovering their solid Parts.[51]

Prescribers were warned to be mindful of dosages, noting that just as chocolate

> produces good Effects, when used moderately, it also ... [produces] bad ones when taken to Excess, or mix't with too many sharp Drugs; for then it causes considerable Fermentations in the Humours, and heats much, and therefore is not good for bilious People. It also hinders People to Sleep, because its exalted Principles cause too great a Rarefaction in the Humours.[52]

Pierre Pomet, chief druggist to Louis XIV, in his *Histoire Générale des Drogues, Simples et Composeés* (1694), also expounded at length upon compounding chocolate concoctions – the finest, he claimed, were available only in Paris. London physician Edward Strother also described the various admixtures containing chocolate purely from a chemical viewpoint. Contemporary London practitioner, John Arbuthnot's account generally conformed with that of Strother, though he also hinted of chocolate's medicinal qualities. For example, the oil of chocolate "seems to be ... rich, alimentary, and anodyne". It is the oil, he emphasised, which gave it special power such that it "often helps Digestion and excites Appetite".[53]

In reviewing the evidence of cacao's appearance in the pharmacopeia literature of later centuries, we find it being mentioned alongside other drugs in such works from a number of countries. Spain listed it in 1739, Britain in 1772, Holland and Germany during the first decade of the 19th century, France the following decade, Sweden and Denmark by 1821, and the United States by

1834. In the latter case, the *Dispensatory of the United States* promoted it, among other uses, as "an excellent substitute for coffee in dyspeptic cases, being nutritive and digestible, without exercising any narcotic or other injurious influence".[54]

Similar to their analysis of the benefits that particular components added to other compounded food and drug products, recipe books and pharmacopeias suggested a variety of ways in which essential ingredients must be combined with chocolate in order to achieve the desired effect. Typical additives to the cacao bean included cinnamon, saffron, nutmeg, Indian or Spanish Pepper, aniseed and vanilla to counteract the bitter tastes; flour made from cassava, maize or Indian corn, each of which acted as an emulsifier, then mixed together with egg yolk to bind it into a paste which was often dried into hard rolls or cakes or bricks. Almonds, hazelnuts and sugar were also frequently added. Such bricks, Dr Strother claimed, consisted of "Particles truly nutritious and alimentary" which, due to the aromatics added, became "very nourishing, cordial and comfortable".[55]

Over time, the health benefits attributed to chocolate tended to increase as rapidly as the growth of chocolate manufacturing.

REFERENCES

1. Though chocolate is known as a bromide (*i.e.*, containing theobromine), bromide was colloquially used in earlier times to mean a sedative or sleep-producing agent, at times also referring to a tiresome or boring person.
2. Before the adoption of its more popular Linnaean classification, the Chocolate Tree was known scientifically as *Amygdalae pecuniariae*, the "money almond".
3. Curiously, it was at the Chicago World's Columbian Exposition in 1893 commemorating the 400th year of Columbus' arrival to the New World that U.S. chocolate magnate Milton S. Hershey experienced something of an epiphany in that chocolate, not caramel, was to be the chief confection of the future. The Exposition opened a year later than the actual commemorative year due to weather delays.
4. Cortés gained special reverence among the Aztecs by appearing at their capital of Tenochtitlán in 1519, the year in which the

god Quetzalcoatl (who had previously developed a following for chocolate drink) was predicted to return.

5. Marcia Morton, *Chocolate: An Illustrated History*, Crown, New York, 1986, p. 6.
6. Sophie D. Coe and Michael D. Coe, *The True History of Chocolate*, Thames and Hudson, London, 1996, pp. 130–131.
7. Carolus Clusius's (Charles de l'Écluse's) 1582 work, which appeared in Latin as noted in the bibliography, recorded *Some Notes on Garcia da Orta's History of Aromatic Plants, Along with Observations on Some Plants and Other Exotic Things Collected by Sir Francis Drake and his Companions on their Voyage Round the World, and on Foreign Products Received by the Author from Friends in London.*
8. Reay Tannahill, *Food in History*, Eyre Methuen, London, 1973, p. 287. See Appendix 2 for the recipe of "Health Chocolate" widely used in Spain.
9. Linda K. Fuller, *Chocolate Fads, Folklore & Fantasies: 1,000 + Chunks of Chocolate Information*, Haworth Press, New York, 1994, p. 8.
10. Linda K. Fuller, *Chocolate Fads, Folklore & Fantasies: 1,000 + Chunks of Chocolate Information*, Haworth Press, New York, 1994, pp. 184–185.
11. M. L. Lémery, *A Treatise of all Sorts of Foods, both Animal and Vegetable: Also of Drinkables: Giving an Account How to Chuse the Best Sort of all Kinds*, translated by D. Hay, W. Innys, T. Longman, and T. Shewell, London, 1745, p. 367.
12. Francesco Carletti, *My Voyage Around the World, Ragionamenti di Francesco Carletti Fiorentino sopra le cose da lui vedute ne' suoi viaggi,* translated from the Italian by Herbert Weinstock, Pantheon Books, New York, 1964, p. 287.
13. Sophie D. Coe and Michael D. Coe, *The True History of Chocolate*, Thames and Hudson, London, 1996, p. 112.
14. Sophie D. Coe and Michael D. Coe, *The True History of Chocolate*, Thames and Hudson, London, 1996, p. 109.
15. Sophie D. Coe and Michael D. Coe, *The True History of Chocolate*, Thames and Hudson, London, 1996, pp. 123–124.
16. Sophie D. Coe and Michael D. Coe, *The True History of Chocolate*, Thames and Hudson, London, 1996, pp. 122–123.
17. For a review of chocolate and humoural medicine, see George M. Foster, *Hippocrates' Latin American Legacy: Humoral*

Medicine in the New World, Gordon and Breach, Langhorne, PA, 1994; Sophie D. Coe and Michael D. Coe, *The True History of Chocolate*, Thames and Hudson, London, 1996, pp. 121–124; Teresa Dillinger, Patricia Barriga, Silvia Escárcega, *et al.*, "Food of the Gods: Cure for humanity? A cultural history of the medicinal and ritual use of chocolate", *JN The Journal of Nutrition*, 2000, **130 Supplement**, pp. 2057S–2072S, pp. 2059S–2060S; Marianna Appel Kunow, *Maya Medicine: Traditional Healing in Yucatán*, University of New Mexico Press, Albuquerque, 2003, pp. 63–65; Meredith L. Dreiss and Sharon Edgar Greenhill, *Chocolate: Pathway to the Gods*, University of Arizona Press, Tucson, 2008, p. 138; and Louis Evan Grivetti, "Medicinal Chocolate in New Spain, Western Europe, and North America", in *Chocolate: History, Culture, and Heritage*, eds. Louis Evan Grivetti and Howard Yana Shapiro, John Wiley & Sons, Hoboken, N. J., 2009, pp. 67–88, p. 69.

18. Sophie D. Coe and Michael D. Coe, *The True History of Chocolate*, Thames and Hudson, London, 1996, p. 123.
19. Linda K. Fuller, *Chocolate Fads, Folklore & Fantasies: 1,000+ Chunks of Chocolate Information*, Haworth Press, New York, 1994, p. 21. Rauch's claim, like that of a Dr Duncan in his 1717 *Wholesome Advice Against the Abuses of Hot Liquors, Particularly of Coffee, Tea, Chocolate, etc.*, seemed to have very little impact upon chocolate consumption. Robin Dand, *The International Cocoa Trade*, 2nd edn, CRC Press, Boca Raton, FL; Woodhead Pub., Cambridge, England, 1999, p. 8.
20. Marcia Morton, *Chocolate: An Illustrated History*, Crown, New York, 1986, p. 20.
21. Diane Barthel discusses chocolate boxes and marketing in "Modernism and marketing: The chocolate box revisited", *Theory, Culture & Society*, 1989, **6**, pp. 429–438.
22. Shara Aaron and Monica Bearden, *Chocolate: A Healthy Passion*, Prometheus Books, Amherst, N. Y., 2008, p. 72.
23. Axel Borg and Adam Siegel, "Early Works on Chocolate: A Checklist", in *Chocolate: History, Culture, and Heritage*, eds. Louis Evan Grivetti and Howard Yana Shapiro, John Wiley & Sons, Hoboken, N. J., 2009, pp. 929–942, pp. 940–941, also cited in G. A. R. Wood and R. A. Lass, *Cocoa*, Longman, London & New York, 1985, p. 5.

24. Devin K. Binder, "The medical history of chocolate", *The Pharos*, 2001, **64**, pp. 22–26, p. 22.
25. H. Stubbe, *The Indian Nectar, or a Discourse Concerning Chocolate wherein the Nature of the Cacao-nut . . . is Examined . . . the Ways of Compounding and Preparing Chocolate are Enquired into; its Effects, as to its Alimental and Venereal Quality, as well as Medicinal (Specially in Hypochondriacal Melancholy) are Fully Debated*, A. Crook, London, 1662, p. 31.
26. H. Stubbe, *The Indian Nectar, or a Discourse Concerning Chocolate wherein the Nature of the Cacao-nut . . . is Examined . . . the Ways of Compounding and Preparing Chocolate are Enquired into; its Effects, as to its Alimental and Venereal Quality, as well as Medicinal (Specially in Hypochondriacal Melancholy) are Fully Debated*, A. Crook, London, 1662, pp. 83–83.
27. H. Stubbe, *The Indian Nectar, or a Discourse Concerning Chocolate wherein the Nature of the Cacao-nut . . . is Examined . . . the Ways of Compounding and Preparing Chocolate are Enquired into; its Effects, as to its Alimental and Venereal Quality, as well as Medicinal (Specially in Hypochondriacal Melancholy) are Fully Debated*, A. Crook, London, 1662, p. 84.
28. Hughes, William, *The American Physitian or A Treatise of the Roots, Plants, Trees, Shrubs, Fruit, Herbs &c. Growing in the English Plantations in America: Describing the Place, Time, Names, Kindes, Temperature, Vertues and Uses of them, either for Diet, Physick, &c. Whereunto is added A Discourse of the Cacao-nut Tree, and the use of its Fruit; with all the ways of making of Chocolate. The like never extant before*, J. C. for William Crook, London, 1672.
29. John Chamberlayne, *The Natural History of Coffee, Thee, Chocolate, Tobacco, in four several Sections; with a Tract of Elder and Juniper-Berries, Shewing how Useful they may be in our Coffee-Houses: And also the way of making Mum, With some Remarks upon that Liquor. Collected from the Writings of the best Physicians, and Modern Travellers*, Christopher Wilkinson, London, 1682, p. 14. John Chamberlayne (1669–1723) was a translator and literary editor of a number of key works. Here, he drew upon many contemporary sources, quoting important works at considerable length offering summative insights into the properties of coffee, chocolate, tea and tobacco.

30. John Chamberlayne, *The Natural History of Coffee, Thee, Chocolate, Tobacco, in four several Sections; with a Tract of Elder and Juniper-Berries, Shewing how Useful they may be in our Coffee-Houses: And also the way of making Mum, With some Remarks upon that Liquor. Collected from the Writings of the best Physicians, and Modern Travellers*, Christopher Wilkinson, London, 1682, pp. 14–15.
31. John Chamberlayne, *The Natural History of Coffee, Thee, Chocolate, Tobacco, in four several Sections; with a Tract of Elder and Juniper-Berries, Shewing how Useful they may be in our Coffee-Houses: And also the way of making Mum, With some Remarks upon that Liquor. Collected from the Writings of the best Physicians, and Modern Travellers*, Christopher Wilkinson, London, 1682, p. 17.
32. John Chamberlayne, *The Natural History of Coffee, Thee, Chocolate, Tobacco, in four several Sections; with a Tract of Elder and Juniper-Berries, Shewing how Useful they may be in our Coffee-Houses: And also the way of making Mum, With some Remarks upon that Liquor. Collected from the Writings of the best Physicians, and Modern Travellers*, Christopher Wilkinson, London, 1682, p. 17.
33. John Chamberlayne, *The Natural History of Coffee, Thee, Chocolate, Tobacco, in four several Sections; with a Tract of Elder and Juniper-Berries, Shewing how Useful they may be in our Coffee-Houses: And also the way of making Mum, With some Remarks upon that Liquor. Collected from the Writings of the best Physicians, and Modern Travellers*, Christopher Wilkinson, London, 1682, p. 17.
34. John Chamberlayne, *The Natural History of Coffee, Thee, Chocolate, Tobacco, in four several Sections; with a Tract of Elder and Juniper-Berries, Shewing how Useful they may be in our Coffee-Houses: And also the way of making Mum, With some Remarks upon that Liquor. Collected from the Writings of the best Physicians, and Modern Travellers*, Christopher Wilkinson, London, 1682, p. 17.
35. John Chamberlayne, *The Natural History of Coffee, Thee, Chocolate, Tobacco, in four several Sections; with a Tract of Elder and Juniper-Berries, Shewing how Useful they may be in our Coffee-Houses: And also the way of making Mum, With some Remarks upon that Liquor. Collected from the Writings of*

the best Physicians, and Modern Travellers, Christopher Wilkinson, London, 1682, p. 17.

36. John Chamberlayne, *The Natural History of Coffee, Thee, Chocolate, Tobacco, in four several Sections; with a Tract of Elder and Juniper-Berries, Shewing how Useful they may be in our Coffee-Houses: And also the way of making Mum, With some Remarks upon that Liquor. Collected from the Writings of the best Physicians, and Modern Travellers*, Christopher Wilkinson, London, 1682, p. 17.
37. William Hughes, *The American Physitian or A Treatise of the Roots, Plants, Trees, Shrubs, Fruit, Herbs &c. Growing in the English Plantations in America: Describing the Place, Time, Names, Kindes, Temperature, Vertues and Uses of them, either for Diet, Physick, &c. Whereunto is added A Discourse of the Cacao-nut Tree, and the use of its Fruit; with all the ways of making of Chocolate. The like never extant before*, J. C. for William Crook, London, 1672, section on "Use" following "Of the Simple Cacao-Kernels".
38. William Hughes, *The American Physitian or A Treatise of the Roots, Plants, Trees, Shrubs, Fruit, Herbs &c. Growing in the English Plantations in America: Describing the Place, Time, Names, Kindes, Temperature, Vertues and Uses of them, either for Diet, Physick, &c. Whereunto is added A Discourse of the Cacao-nut Tree, and the use of its Fruit; with all the ways of making of Chocolate. The like never extant before*, J. C. for William Crook, London, 1672, section on "Name" in the chapter "Of the Cacao-Tree and Fruit".
39. William Hughes, *The American Physitian or A Treatise of the Roots, Plants, Trees, Shrubs, Fruit, Herbs &c. Growing in the English Plantations in America: Describing the Place, Time, Names, Kindes, Temperature, Vertues and Uses of them, either for Diet, Physick, &c. Whereunto is added A Discourse of the Cacao-nut Tree, and the use of its Fruit; with all the ways of making of Chocolate. The like never extant before*, J. C. for William Crook, London, 1672, following "To The Reader".
40. Philippe Sylvestre Dufour, *De l'Usage du Caphé, du Thé et du Chocolate*, Iean Girin & Barthelemy Riviere, Lyon, 1671, was translated into, among other languages, English as P.S. Dufour, *The Manner of Making Coffee, Tea and Chocolate as it*

is Used in Most Parts of Europe, Asia, Africa and America, With their Vertues, W. Crook, London, 1685.

41. P. S. Dufour, *The Manner of Making Coffee, Tea and Chocolate as it is Used in Most Parts of Europe, Asia, Africa and America, With their Vertues*, W. Crook, London, 1685, pp. 115–116.
42. Sophie D. Coe and Michael D. Coe, *The True History of Chocolate*, Thames and Hudson, London, 1996, p. 156. René Moreau dedicated his *Du Chocolat Discours Curieux* (1643) – a French translation of Colmenero de Ledesma's work – to the Cardinal of Lyon.
43. Marcia Morton, *Chocolate: An Illustrated History*, Crown, New York, 1986, p. 20.
44. Christine McFadden and Christine France, *Chocolate: Cooking with the World's Best Ingredient*, Hermes House, New York, 2001, p. 22. See also Samuel Pepys' *Diary*, 1661, Wednesday 24 April, http://www.pepysdiary.com/archive/1661/04, accessed 31 May 2011.
45. Marcia Morton, *Chocolate: An Illustrated History*, Crown, New York, 1986, p. 28. According to Robert Whymper, *Cocoa and Chocolate: Their Chemistry and Manufacture*, J. & A. Churchill, London, 1912, p. 8, the supervised growth of Chocolate Trees in French Martinique, which began in 1679, represented the "first step" in the "organized cultivation of cacao".
46. M. Rabutin-Chantal, *Letters choisies*, Garnier Frères, Paris, 1878. Letters 11.2.1671, 15.4.1671, 13.5.1671, 23.10.1671.
47. Ambergrese (or Ambergris) was the gray amber looking, pleasant floral smelling waxy accretion formed in the gastrointestinal track of sperm whales that, after being vomited up, became a highly prised delicacy to humans who used it for, among other things, thickening their chocolate drinks. René Moreau noted this use of ambergris in chocolate preparation in his *Du Chocolat: Discours Curieux Divisé en Quatre Parties* – Sebastien Cramoisy, Paris, 1643. For further discussion of whaling and the chocolate industry, see Christopher Kelly, "Chocolate and North American Whaling Voyages", in *Chocolate: History, Culture, and Heritage*, eds. Louis Evan Grivetti and Howard Yana Shapiro, John Wiley & Sons, Hoboken, N. J., 2009, pp. 413–424.
48. John Chamberlayne, *The Natural History of Coffee, Thee, Chocolate, Tobacco, in four several Sections; with a Tract of*

Elder and Juniper-Berries, Shewing how Useful they may be in our Coffee-Houses: And also the way of making Mum, With some Remarks upon that Liquor. Collected from the Writings of the best Physicians, and Modern Travellers, Christopher Wilkinson, London, 1682, pp. 15–16.

49. M. L. Lémery, *A Treatise of all Sorts of Foods, both Animal and Vegetable: Also of Drinkables: Giving an Account How to Chuse the Best Sort of all Kinds*, translated by D. Hay, W. Innys, T. Longman, and T. Shewell, London, 1745, p. 364.
50. M. L. Lémery, *A Treatise of all Sorts of Foods, both Animal and Vegetable: Also of Drinkables: Giving an Account How to Chuse the Best Sort of all Kinds*, translated by D. Hay, W. Innys, T. Longman, and T. Shewell, London, 1745, pp. 365–366. Bezoar stones, typically used as antidotes to poisons, were clumps or concretions of indigestible masses of material taken from the stomachs or intestines of ruminants. French refinements in chocolate preparation are featured in a number of works, many of which are cited in Nikita Harwich's *Histoire du Chocolat* – Editions Desjonquères, Paris, 1992.
51. M. L. Lémery, *A Treatise of all Sorts of Foods, both Animal and Vegetable: Also of Drinkables: Giving an Account How to Chuse the Best Sort of all Kinds*, translated by D. Hay, W. Innys, T. Longman, and T. Shewell, London, 1745, p. 366.
52. M. L. Lémery, *A Treatise of all Sorts of Foods, both Animal and Vegetable: Also of Drinkables: Giving an Account How to Chuse the Best Sort of all Kinds*, translated by D. Hay, W. Innys, T. Longman, and T. Shewell, London, 1745, p. 366.
53. John Arbuthnot, *An Essay Concerning the Nature of Ailments*, 2nd edn, J. Tonson, London, 1732, p. 149.
54. G. B. Wood and F. Bache, eds., *Dispensatory of the United States*, Gregg and Elliot, Philadelphia, 2nd edn. 1834, pp. 1079–1080. Martha Makra Graziano makes excellent use of evidence drawn from pharmacopeias in her "Food of the Gods as mortal's medicine: The uses of chocolate and cacao products", *Pharmacy in History*, 1998, **40**, pp. 132–146.
55. Edward Strother, *Materia Medica: or, a New Description of the Virtues and Effects of all Drugs, or Simple Medicines now in Use*, Charles Rivington, London, 1727, vol. 2, p. 16.

CHAPTER 4

Expanding Chocolate's Use as Medicine

> [Physicians were] the first to mix various chocolate preparations. This was because they were considered to be adept at making remedies and chocolate was considered a 'confection' – which meant, during that era, a medicine which could be made to taste good only by mixing it with sugar or other good-tasting substances. It was only natural that doctors should be among the pioneer manufacturers [of chocolate].
>
> P.P. Gott and L.F. Van Houten,
> *All About Candy and Chocolate* (1958)

Much has been written on chocolate, its myriad recipes and its consumption during the 18th century. Indeed, there was "hardly an illness [during this period] for which chocolate was not presented as a cure, scarcely a sensory function whose improvement was not attributed to chocolate, and not a few of the leading figures of the time [who] freely admitted that they owed a great deal to chocolate".[1] In this chapter, we shall first examine an array of the reputed medicinal uses of chocolate in Europe as well as in Colonial America, then we shall focus upon the novel method of preparing and consuming chocolate that ultimately led to an entire new industry.

Chocolate as Medicine: A Quest over the Centuries
Philip K. Wilson and W. Jeffrey Hurst

Published by the Royal Society of Chemistry, www.rsc.org

4.1 MEDICS AND MANUFACTURERS PROMOTING CHOCOLATE

Country by country, chocolate became touted as a medicinal remedy throughout Europe. In the form of a paste, it was exported from Spain to Italy and France. Spanish Court Physician Augustin Farfan referenced chocolate's ability to eliminate kidney stones and purge the gut in his 1579 *Tractado Breve de Medicina*. The *Un Discurso del Chocolate* (1624) elaborated upon New World views of chocolate's benefit as a medicine for Old World audiences. Francisco Maria Brancaccio, later Cardinal Brancaccio, described chocolate's usefulness as a medicine that "restores natural heat, generates pure blood, enlivens the heart, [and] conserves the natural faculties".[2] René Moreau prepared a medical dissertation on the healthfulness of chocolate in 1643, and by 1659, the Paris Faculty of Medicine had bestowed their imprimatur on its use (Figure 4.1). Cornelis Bontekoe, Dutch physician to the Elector Wilhelm of Brandenburg, increased the consumption of chocolate (as well as tea) in Germany through his favorable review of its medical qualities in his 1678 *Tractaat van het Excellenste Kruyd Thee*. In 1706, Daniel Duncan composed a treatise, *Wholesome Advice Against the Abuse of Hot Liquors, Particularly of Coffee, Tea, Chocolate*, noting chocolate to be healthy only when drunk in moderation. The uses as well as the abuses of chocolate were examined in Leonhard Ferdinand Meisner's 1721 *Caffe, Chocolatae, Herbae Thee ac Nicotianae: Natura, Usu, et Abusu Anacrisis: Medico-Historico-Diatetica* as well as in Girolamo Giuntini's *Altro Parere Intorno alla Natura ed all'uso della Cioccolata Disteso* (1728) (Figure 4.2).[3] Remedies and recommendations for chocolate's use in fighting disease abounded in François Foucault's 1684 medical dissertation (prepared under the supervision of Stèphane Bachot), *An Chocolatae Usus Salubris?*, in D. Quélus' 1719 publication, *Histoire Naturelle du Cacao et du Sucre*, in Pierré Joseph Buc'hoz's *Dissertations sur l'Utilité, et les Bon et Mauvais Effects du Tabac, du Café, du Cacao, et du Thé* (1788), and in Munster College of Medicine Director, Christopher Ludwig Hoffmann's late 18th-century treatise, *Potus Chocolate*.

The entrepreneurial English physician William Salmon (who some contemporaries dubbed a quack)[4] prepared a special "liquid medicine" he named "chocolate wine" which he prescribed and

DV
CHOCOLATE.
DISCOVRS CVRIEVX,
DIVISÉ EN QVATRE PARTIES.
Par Antoine Colmenero de Ledeſma Medecin & Chirurgien de la ville de Ecija de l'AndalouZie.
Traduit d'Eſpagnol en François ſur l'impreſſion faite à Madrid l'an 1631. & eſclaircy de quelques Annotations.
Par RENE' MOREAV *Profeſſeur du Roy en Medecine à Paris.*
Plus eſt adjouſté vn Dialogue touchant le meſme Chocolate.
Dedié à Monſeigneur l'Eminentiſſime Cardinal de Lyon, grand Aumoſnier de France.

A PARIS.
Chez SEBASTIEN CRAMOISY, Imprimeur ordinaire du Roy, ruë S. Iacques, aux Cicognes.
M. DC. XLIII.

Figure 4.1 Title page of René Moreau's *Medical Discourse on the Healthfulness of Chocolate* (1643).
(Courtesy of Hershey Community Archives, Hershey, Pennsylvania, USA).

LEONH. FERDINAND. MEISNERI,
Med. Doct. & Prof. Regii,
DE
CAFFE, CHO-
COLATÆ, HERBÆ
THEE ac NICOTIANÆ
NATURA, USU, ET ABUSU
ANACRISIS,
Medico-Hiſtorico-
Diætetica.

NORIMBERGÆ,
Sumpt. JOH. FRIDER. RUDIGERI,
ANNO MDCCXXI.

Figure 4.2 Title page of Leonhard Ferdinand Meisner's Work on the Healthfulness of Chocolate and Related Natural Products (1721). (Courtesy of Hershey Community Archives, Hershey, Pennsylvania, USA).

sold to his patients to be drunk by the glassful. William Buchan, author of the best-selling *Domestic Physician*, deemed chocolate a preventative for "fainting fits", whereas his contemporary, the polymath and industrial revolutionary, Dr Erasmus Darwin promoted chocolate's benefits based in part upon his own testimonial of its help in treating gout. Linnaeus, who is often noted only as the nosologist who named the Chocolate Tree as *Theobroma cacao* in his *Species Plantarum* (1753), also noted his findings of its medicinal benefits in his Swedish treatise, *Om Chokladdrynen* (1778). In particular, Linnaeus promoted its use for three classes of disorders: wasting of the body from diseases of the lungs or muscles, hypochondria, and hemorrhoids. His contemporary, the physician from St Dizier, Champagne, Pierre-Toussaint Navier, also spoke of chocolate as medicine in *Observationes sur le Cocao et sur le Chocolat* (1772, and in German translation, 1775). Among the benefits Navier noted were chocolate's usefulness against scurvy, consumption, worms, digestive acids and general disorders of the lungs, heart and vessels. In details far beyond the typical descriptions of the day, Navier articulated his view that chocolate's particular usefulness to gut and bowel disorders was due to its being "incorruptible" throughout the digestive tract. Unlike milk and meat which experimenters of the day had found to easily go rancid, chocolate's cocoa butter content offered it a "high degree of resistance to rancidity".[5] For those who were unable to "stomach" cacao's high fat concentrations, Navier recommended cacao shell infusions that he claimed, in medical terms, to be "stomachic, balsamic, pectoral, and especially aperient" in its properties. In addition to noting cacao's own medical uses, Navier further described chocolate's benefit as a vehicle for other types of medicines such as purgatives, attenuants, expectorants, diuretics and incidentia.[6]

In addition to physicians prescribing chocolate, its healthy benefits were widely discussed in Chocolate Houses – the ever growing "Penny Universities" to which consumers flocked in cosmopolitan centers, particularly London, during the early 1700s. For a penny entrance fee, one could have access to newspapers of the day – both in print form and read aloud – as well as to general city scuttlebutt.[7] There in the press locals were reminded that Cortés had once claimed that a "cup of this precious drink [chocolate] permits a man to walk for a whole day without food".[8] His claim paralleled Johann Wolfgang von Goethe's own later notion that,

"Whoever has drunk a cup of chocolate can endure a whole day's travel".[9]

In these centers of public learning, chocolate was consumed in great quantities. Waiters were readily available and responded to the incentive of leaving a coin in a brass wall box marked "To Insure Promptness", the origin of our colloquial term, "TIP".[10] Eventually, people spent such vast amounts of time consuming chocolate that they inquired about meals being served in Chocolate Houses, following which the concept of restaurants originated.

Chocolate Houses were also the home of journalistic writing, some of which during the early 18th century was directed to chocolate's potential medical effects. On Chesterfield Street, Mrs White's Chocolate House (established by the Italian immigrant, Mr Francesco Bianco) became a fashionable and civilised London meeting place for members of Whig politics to sip healthy and wholesome chocolate. The Cocoa Tree Club, a Chocolate House at 64 St James Street, Piccadilly, London that had initially served as Jacobite headquarters, then an anti-Whig Tory political club, and later a literary club, had become the space in which Richard Steele and Joseph Addison (both impassioned consumers of chocolate) composed their articles for the daily newspaper, *The Spectator*, during 1711 and 1712. This newspaper's stated aim was "to enliven morality with wit, and to temper wit with morality ... to bring philosophy out of the closets and libraries, schools and colleges, to dwell in clubs and assemblies, at tea-tables and coffeehouses". Addison referenced the increasing use of chocolate in England in 1711 stating, "we repair our bodies by the drugs of America".[11] A year later, a *Spectator* entry warned readers to be "careful" how they "meddle with romances, chocolate, novels and the like inflamers [of passion]".[12] One can readily imagine that it was in such establishments where contemporaries learned that the chocolate consumed therein was heralded as holding

> every "virtue proper to remedy ... [the] inconveniences" of aging. The volatile sulphur with which [chocolate] abounds is proper to supply the place of that which the blood loses every day through age; it blunts and sheaths the points of the salts and restores the usual softness to the blood .... The same sulphurous unctuosity ... spreads itself in the solid parts and gives them ... their natural suppleness; it bestows on the

> membranes, the tendon, the ligaments and the cartilages a kind of oil which renders them smooth and flexible. Thus the equilibrium between the fluids and the solids is, in some measure, re-established, the wheels and springs of our machine mended, health is preserved and life prolonged.[13]

It was in a Belgium Chocolate House in 1697 where the visiting Zurich Burgomaster, Henri Escher, first tasted chocolate, and he subsequently introduced that drink back into his native Switzerland. Although Chocolate Houses slowly faded away by the 1800s, the demand for chocolate drinks did not. Their supply was offered at another venue that escalated in popularity during the 18th and early 19th centuries – spa towns.[14] Chocolate drinks were frequently offered to the convalescents or invalids who resorted to spa towns in order to "take the waters" for their respective complaints. Common throughout European spa towns, chocolate was introduced in the North American spa town of Saratoga Springs, New York in 1791.[15]

Dispensing chocolate as medicine was not without constraints. Various regulatory bodies took charge of chocolate distribution within certain regions. Vienna's chocolate makers came together to form a guild in 1744 to officially recognise the importance of the various products they made "for pleasure drinking and for medicinal use".[16] The Prussian and Hessian German States deemed the need to add special taxes upon chocolate's import and use. There, as in England, import tax records confirm that, at least initially, these countries received processed (*i.e.*, not raw) cacao from Spain and France. Excise men supplied stamped papers in which chocolate bricks, blocks or sticks were wrapped and then sealed, thereby proving that the taxes had been paid.[17] U.S. Secretary of the Treasury Alexander Hamilton established tariffs on chocolate as well as on sugar in 1792. At times, these taxes and tariffs were offset by classifying chocolate solely as a drug in locales where everything other than medicine was taxable.[18] Legal case studies indicate that considerable adulteration took place in the processing and packaging of what was sold under the name of chocolate. Though such taxes as well as rumors of counterfeit chocolate somewhat stifled its use in these regions during the early 18th century, medical authorities of Germany and elsewhere continued to "welcome it with laudatory treatises".[19]

At times and in certain places, moving chocolate from country to country was, indeed, taxing. At others, special circumstances underscored such transnational moves. Consider, for instance, that Marie Antionette (whom the French disdainfully referred to as "The Austrian"), the daughter of Vienna's Emperor Francis I and Empress Maria Theresa, easily transported chocolate via her personal chocolatier whom she took with her to Paris and who became known as "Chocolate Maker to the Queen".[20] Part of his duties included making medicine, chocolate medicine in particular. His medicines for the Queen popularised the use of adding flavorings to chocolate. Powdered orchid bulbs mixed with chocolate were found to be helpful for "plumping out the figure charmingly", orange blossoms were added to chocolate preparations "to soothe the nerves", and the "milk of almonds" was mixed with chocolate when used to "bolster a delicate stomach".[21]

Stephani Blancardi, physician in the bustling chocolate port city of Amsterdam, wrote of chocolate in 1705 as a "veritable balm of the mouth" which was useful for "maintaining all of the glands and humours in a good state of health". But beyond having long appreciated chocolate's medical usefulness, the Dutch also envisioned chocolate's great sales potential. Indeed, it was the Dutch in New Amsterdam (later New York) who reintroduced chocolate to the Americas in the 17th century, this time to the northern continent. Chocolate had been a New World substance that had been introduced to the Old World together with maize, potatoes, peanuts, vanilla, tomatoes, pineapples, "French" beans, Lima beans, scarlet runners, red and green peppers, tapioca and turkey. In due course, chocolate was introduced to North America via traders of the Old World – together with useful substances including gold, silver, tobacco, rubber and quinine.[22]

By 1712, a Boston apothecary was advertising chocolate in his store. Chocolate became regularly listed on prescriptions written throughout the 18th century for the prevention and care of smallpox. In Colonial America, threats of smallpox epidemics produced a scare second only to the plague. During this time, almanacs served as a popular guide to the use of chocolate regarding smallpox. Benjamin Franklin's *Poor Richard's Almanack* [sic] of 1761 included chocolate as a recommended beverage ingredient when "illing" with smallpox. Following the Boston smallpox epidemic of 1764, *Hutchin's Improved Almanack* [sic] recommended chocolate as part of the preparation for smallpox inoculations.[23] During the 1780s, Philadelphia's "Revolutionary Physician" Benjamin Rush prescribed weak chocolate with a biscuit as part of preparing his patients for smallpox inoculations and, in 1788, he writes of having regularly prescribed chocolate as part of the diet for the many smallpox patients he treated.[24]

Near Boston a half century later, in 1764, the disgruntled Harvard educated minister-turned-physician and teacher, Dr James Baker, employed the Irish immigrant and chocolate maker, John Hannon. This venture ultimately modified the way in which chocolate was prepared, sold and used in Colonial America. Baker and Hannon ground cacao beans in a water-powered grist mill they established at the Lower Mills of Milton, the site of an earlier powder mill on the Neponset River (later known as the "river of American business") ten miles from Boston, across the river from Dorchester.

Baker helped finance the operation of what became known as the John Hannon Company, the first long-term producer of chocolate in Colonial America.[25] Typical of the era, they advertised that their bricks (or hard cakes) of ground and molded chocolate powder should, after being mixed with boiling water, be used "primarily for ... medicinal values".[26] Their product, sold out of a General Store within Baker's Dorchester home, was marketed with great assurance, each label noting that, "If the Chocolate does not prove good, the Money will be returned". In 1779, Hannon was lost at sea amidst the throng of British naval ships along the Colonial eastern seaboard while enroute to the West Indies to purchase more cacao beans. Baker took over the company in 1780 and, though since 1989 it has been owned by Kraft Family Foods and has been relocated to Dover, Delaware, the baking chocolate they produce still bears the brand name Baker's Chocolate (Figure 4.3).[27]

America's Thomas Jefferson claimed, in 1785, that the "superiority of chocolate [drink], both for health and nourishment, will soon give it the same preference over tea and coffee in America which it has in Spain".[28] Powerful though such claims were, the prevalence of this drink throughout the Old World and North America did not become prominent for many years. Part of this time lag may be accounted for by the consistency of chocolate drink at this time.

Figure 4.3 Doctors in 1840 Recommend Baker's Chocolate (1900).

The Colonial drink produced by Baker had much more of a fatty consistency than what we drink today. Refinement of the process to remove substantially more cocoa butter from the powder, part of drinking chocolate's real "coming-of-age", did not occur until 1828. It was then that Holland's Coenraad Johannes Van Houten, son of chocolate miller, Casparus Van Houten, refined cacao into a much more palatable and digestible form by extracting the natural fat (cocoa butter) from the bean, leaving only compressed hard cakes of the cacao nib (or cotyledon) that were pulverised into powder. This reduced fat powder was then mixed with potash to darken its color, lighten its flavor and improve its solubility.[29] The powder produced by this "Dutching" process is what has become widely known as cocoa. Never far removed from the use of chocolate as medicine, Van Houten's mixture soon became advertised as "The Food Prescribed by Doctors" (Figure 4.4).

4.2 MILK IS ADDED TO CHOCOLATE

One noticeable difference in the use of chocolate during the early 1700s from that of earlier periods was the way in which it became both figuratively and literally linked with milk. Today, it is the Swiss nation that has become synonymous with milk chocolate. Their special modification of milk chocolate stems from Henri Nestlé's 1867 preparation of condensed powdered milk by evaporation that forever changed the taste of Swiss chocolate. Yet milk chocolate's heritage extends even further back in time – and it is tied precisely to a medicinal use (Figure 4.5).

London physician and Royal Society President Sir Hans Sloane specifically touted his Milk Chocolate drink as the new restorative – an additive to his growing medical armamentarium gleaned from his 1687 voyage to Jamaica. Following the end of the Anglo-Dutch War in 1654, Lord Protector Cromwell's forces invaded Spain's territories in the West Indies, initially resulting in the capture of Jamaica from Spain in 1655 under the joint command of British Army Lieutenant-Colonel Robert Venebles and British Navy Admiral Sir William Penn whose son gained note in 1681 as founder of the American Province (or Colony) of that future chocolate haven, Pennsylvania. Upon the British takeover of Jamaica, this land's natural resources, in particular the sugar plantations, were deemed to be of great value.

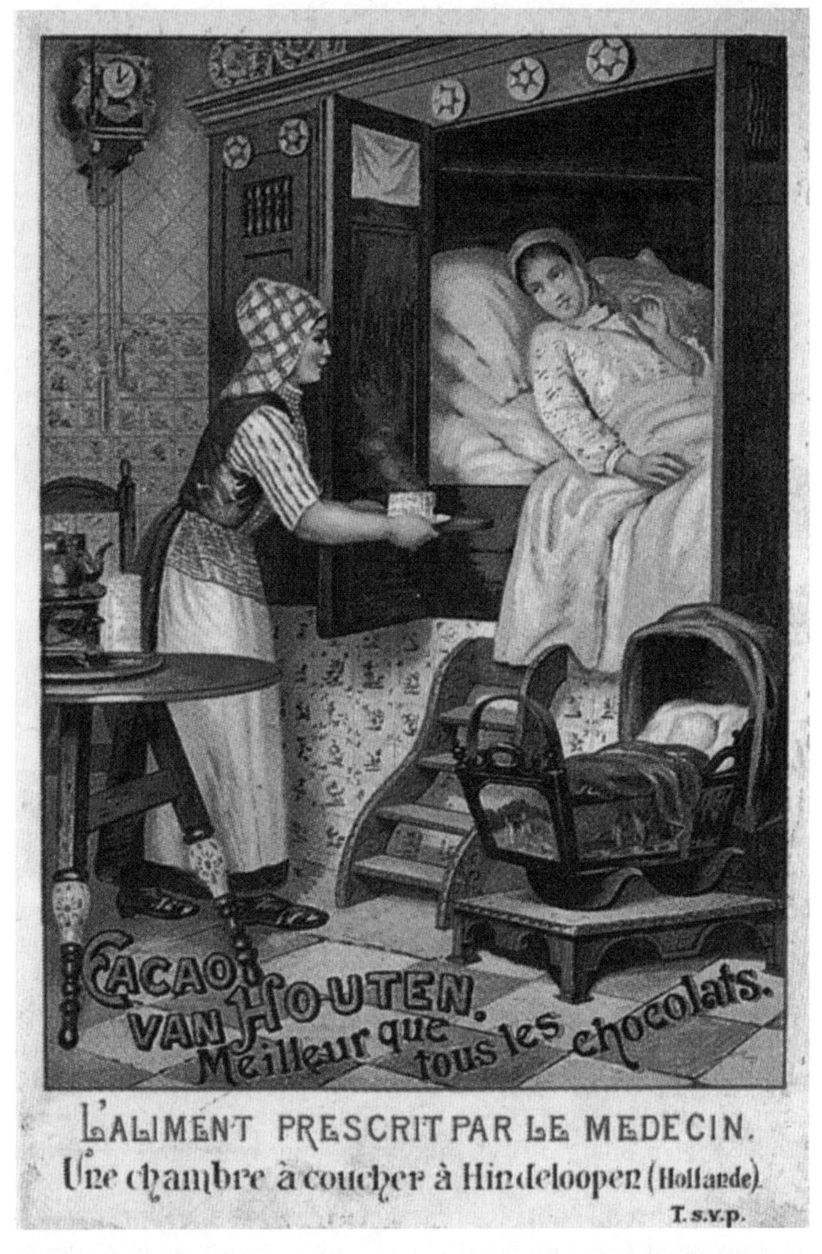

Figure 4.4 Van Houten's 19th-Century Trade Card. Chocolate – The Food Prescribed By Doctors.

Figure 4.5 Nestlé's Milk Food Label (1875).

Sloane had undertaken his 1687 voyage as physician to the new Governor of Jamaica, Christopher Monck, 2nd Duke of Albemarle. During his fifteen months in Jamaica, Sloane took copious notes and extracted ample specimens of local flora and fauna as well as ethnographically noting information about the indigenous human population. Following the Duke's death, Sloane returned to England in 1689 and gathered information from the voyage into a massive and intricate collection of botanical observations and knowledge, the *Catalogus Plantarum*, which he published in 1696. Sloane published two more extensive volumes in the 18th century based upon the information collected during his travels of the 1680s as *Voyage to the Islands Madera, Barbados, Nieves, S. Christophers and Jamaica, with the Natural History of the Herbs and Trees, Four-footed Beasts, Fishes, Birds, Insects, Reptiles, &c. Of the last of those Islands*. His collecting interest that was initiated on this voyage approached a mania, resulting in a massive group of artifacts that ultimately filled his Bloomsbury Square residence which soon became known as Sloane's Museum. With the hope that his vast collections "may remain together and not be separated" and stay in London, "where I have acquired most of my estates and where they may by the great confluence of people be most used", he offered his collection, upon his death, to Great Britain.[30] The required sum of £20 000 was raised by lottery, and his collections became the nidus for both the

Figure 4.6 Sir Hans Sloane Recognized as Founder of the British Museum. (Personal Collection).

British Museum (Bloomsbury) and later, The Natural History Museum (South Kensington) (Figure 4.6).

Among Sloane's treasured finds in Jamaica was the chocolate drink that the natives enjoyed. Later in London, Sloane bestowed the "secret" remedy to Nicholas Sanders and William White who manufactured the product and advertised its benefits as a medicinal drink that produced a "Lightness on the Stomach" yet proved to be of "great Use in all Consumptive Cases".[31] In his own writings, Sloane noted chocolate's additional benefits of "dissipat[ing] malignant Humours" from settling on the breasts, carrying off gravel and stones of the kidneys, diminishing bloody stools, and halting the progress of continued or remittent fevers which would, without chocolate, lead to emaciation.[32] Adding milk to his chocolate concoction was, for Sloane, a method to improve its digestibility. Sloane was not the only London merchant of "milk chocolate" for by 1659, it was being sold at, among other venues, Queen's-Head Alley off Bishopsgate Street as a medicine that "cures and preserves the body of many diseases".[33]

Before chocolate was known in Europe, wine was called the "milk of old men", but by the early 1700s, contemporaries noted that this title was "now [being] applied with greater reason to chocolate, since its use has become so common, that it has been perceived that chocolate is, with respect to them, what milk is to infants".[34] This claim derived primarily from Sloane's recipe, the medical document that introduced and first popularised the product "milk chocolate".

John Cadbury, together with sons George and Richard, later purchased Sloane's Milk Chocolate recipe and prepared it in their factory along the river Bourn. The Cadburys were Quakers who, like the Frys, Tukes, Rowntrees and a number of other families of this dissenting denomination, came to view chocolate as a nourishing, healthy alternative to alcohol. The Cadburys specifically promoted it as a healthy "flesh forming" substance.[35] In order to further enhance its popularity, the Cadbury Brothers advertised their special Sloane remedy between 1849 and 1885 as a "health food", adding the rhetorical claim that to call it a "medication would not be too strong a term".[36]

Contemporary French manufacturers were also touting their own milk chocolate remedies as being specifically beneficial for individuals with fragile stomachs as well as more generally for convalescents and children. In 1800, French pharmacist to King Louis XVI, Sulpice Debauve (and after 1823, with Antoine Gallais) gained renown for a special "medical chocolate" product.[37] By this time, milk chocolate's virtues were being touted both as relieving specific disorders and as a universal cure-all.

By diluting chocolate's singular nature with the addition of milk, its once primary use as a medicine began to murkily blur with another potential primary use as a food. The dual role of chocolate as both medicine and food carried over into the professional literature as well as in the popular press. The 1834 *Dispensatory of the United States*, for example, touted the cocoa commonly served "as a drink at the morning and evening meals" to also be "an excellent substitute for coffee in dyspeptic cases, being nutritive and digestible, without exercising any narcotic or other injurious influence". The same entry also promoted cocoa as "a good article of diet for convalescents" that "may sometimes be given advantageously as a mild nutritive drink in cases of disease".[38] As the next chapter will demonstrate, this dual approach to advertising chocolate's benefits truly began to flourish throughout the 19th century.

REFERENCES

1. Chocosuisse, the Union of Swiss Chocolate Manufacturers' *Chocologie*, p. 6, as cited by Linda K. Fuller, *Chocolate Fads, Folklore & Fantasies: 1,000+ Chunks of Chocolate Information*, Haworth Press, New York, 1994, p. 10.

2. Sophie D. Coe and Michael D. Coe, *The True History of Chocolate*, Thames and Hudson, London, 1996, p. 154.
3. Additional works of this period discussing chocolate's medical potential include Giovanni Batista Felici, *Parere Intorno all'uso della Cioccolata: Scritto in una Lettera*, Appresso G. Manni, Florence, Italy, 1728, and Giuseppe Demarco, *Josephi Demarco Medici Melitensis Philosophiæ, & Universitatis Monspelliensis medicinæ doctoris De Lana Rité in Secunda, & Adversa valetudine adhibenda: Opus, quo Villosæ Vestis nudi contactus Præstantia, &c Actio Staticæ experimentis perspicué, Utilitates fusé demonstrantur, Noxæ diligenter expenduntur. Adjecta est ad calcem Dissertatio de usu, et abusu chocolatæ in Re Medica, & Morali*, Melitæ: in Palatio, & ex Typographia C. S. S. Apud D. Nicolaum Capacium ejus Typographum, 1759.
4. Philip K. Wilson, "William Salmon", in *New Dictionary of National Biography*, ed. H. C. G. Matthew, Oxford University Press, Oxford, 2004, **48**, pp. 734–735. For more on quacks of the period, see Eric Jameson, *The Natural History of Quackery*, Michael Joseph, London, 1961, Roy Porter, *Health for Sale: Quackery in England, 1660–1850*, Manchester University Press, Manchester, 1989, and Philip K. Wilson, *Surgery, Skin & Syphilis: Daniel Turner's London (1667–1741)*, Wellcome Institute Series in the History of Medicine, Clio Medica **54**, Rodopi Press, Amsterdam and Atlanta, 1999, esp. pp. 91–99.
5. Pierre Toussaint Navier, *Bemerkungen über den Cacao und die Chocolate, worinnen der Nutzen und Schaden untersuchet wird, der aus dem Genusse dieser nahrhaften Dinge entsehen kann: Alles auf Erfahrung und zergliedernde Versuche mit der Cacao-Mandel gebauet; Nebst einigen Erinnerungen über das System des Hrn. De-La-Müre, betreffend das Schlagen der Puls-adern*, Saalbach, Leipzieg, 1775, p. 95.
6. Martha Makra Graziano, "Food of the Gods as mortals' medicine: The uses of chocolate and cacao products", *Pharmacy in History*, 1998, **40**, pp. 132–146, p. 136.
7. For an historical overview of coffee houses, see Aytoun Ellis, *The Penny Universities: A History of the Coffee-Houses*, Secker & Warburg, London, 1956, and Markman Ellis, ed., *Eighteenth-Century Coffee-House Culture*, Pickering & Chatto, London, 2006.

8. Marcia Morton, *Chocolate: An Illustrated History*, Crown, New York, 1986, p. 6.
9. Carole Bloom, *All About Chocolate: the Ultimate Resource to the World's Favorite Food*, Macmillan, New York, 1998, p. 165.
10. Norah Smaridge, *The World of Chocolate*, J. Messner, New York, 1969, pp. 26–27.
11. C. Trevor Williams, *Chocolate and Confectionary*, Leonard Hill, London, 1950, 3rd edn, 1964, p. 7.
12. Robert Whymper, *Cocoa and Chocolate: Their Chemistry and Manufacture*, J. & A. Churchill, London, 1912, p. 7.
13. Written in the reign of Queen Anne (1702–1714), as cited in "In Praise of Chocolate", *Confectioners' Journal*, 1907, **33**, p. 94.
14. Robin Dand, *The International Cocoa Trade*, 2nd edn, CRC Press, Boca Raton, FL; Woodhead Pub., Cambridge, England, 1999, p. 7.
15. By the early 20th century, it was "still customary at some hydropathic establishments ... for doctors to order [cacao] 'nibs' for their patients", this act being a "despairing effort to obtain the genuine article", Brandon Head, *The Food of the Gods: A Popular Account of Cocoa*, George Routledge & Sons, London, E. P. Dutton, New York, 1903, pp. 17–18.
16. Marcia Morton, *Chocolate: An Illustrated History*, Crown, New York, 1986, p. 67.
17. Chantal Coady, *Chocolate: The Food of the Gods*, Chronicle Books, San Francisco, 1993, p. 38.
18. David Lebovitz, *The Great Book of Chocolate: The Chocolate Lover's Guide, with Recipes*, Ten Speed Press, Berkeley, 2004, p. 40.
19. Marcia Morton, *Chocolate: An Illustrated History*, Crown, New York, 1986, p. 27.
20. Empress Maria Theresa (of Spanish descent as daughter of Hapsburg Emperor Charles VI who had been previously crowned Charles III of Spain) was a noted "chocoholic" who encouraged the Viennese to enjoy this Spanish delight widely without imposing an import tax. She also gained public notice for the "Cantata to Chocolate" that she commissioned from her court poet.
21. Marcia Morton, *Chocolate: An Illustrated History*, Crown, New York, 1986, p. 41.

22. Reay Tannahill, *Food in History*, Eyre Methuen, London, 1973, pp. 241, 263.
23. Louis Evan Grivetti, "Chocolate and the Boston Smallpox Epidemic of 1764", in *Chocolate: History, Culture, and Heritage*, eds. Louis Evan Grivetti and Howard Yana Shapiro, John Wiley & Sons, Hoboken, N. J., 2009, pp. 89–98, pp. 90–91.
24. Deanna L. Pucciarelli and Louis E. Grivetti, "The medicinal use of chocolate in early North America", *Molecular Nutrition & Food Research*, 2008, **52**, pp. 1215–1227, pp. 1219–1220.
25. In 1752, Obadiah Brown had established a water-powered mill in Colonial Providence, Rhode Island, which produced chocolate for Newport-based merchants, though Brown's business was short lived.
26. Julie Pech, *The Chocolate Therapist: Chocolate Remedies for a World of Ailments*, Trafford, Victoria, B.C., 2005, p. 20.
27. Baker's earlier product, however, tasted somewhat different for although they had ground out much of the cocoa butter from the beans, that butter was still used in preparing the paste that was molded and hardened into their bricks of chocolate. For a more complete history, see Anthony M. Sammarco, *The Baker Chocolate Company: A Sweet History*, History Press, Charleston, SC, 2009.
28. Marcia Morton, *Chocolate: An Illustrated History*, Crown, New York, 1986, p. 34.
29. Bernard W. Minifie, *Chocolate, Cocoa and Confectionery: Science and Technology*, 3rd edn, Van Nostrand Reinhold, New York, 1989, p. 3, claimed that this process produced a powder with only 23% fat.
30. Sir Hans Sloane's will, as cited by E. St John Brooks, *Sir Hans Sloane: The Great Collector and his Circle*, Batchworth Press, London, 1954, p. 219. For more on Sloane, his museum and his use of chocolate, see Arthur MacGregor, ed., *Sir Hans Sloane: Collector, Scientist, Antiquary, Founding Father of the British Museum*, British Museum Press, London, 1994.
31. For more on Sloane and milk chocolate, see T. B. Rogers, *A Century of Progress, 1831–1931: Cadbury, Bourneville*, [Hudson & Kearns, London, 1931], pp. 20–21, and Harold McGee, *On Food and Cooking: The Science and Lore of the Kitchen*, Charles Scribner's Sons, New York, 1984, pp. 399–400.

32. Martha Makra Graziano, "Food of the Gods as mortals' medicine: The uses of chocolate and cacao products", *Pharmacy in History*, 1998, **40**, pp. 131–146, p. 136.
33. Sophie D. Cole and Michael D. Cole, *The True History of Chocolate*, Thames and Hudson, London, 1996, p. 136. Bertha S. Dodge also mentions this in her *Plants that Changed the World*, Little, Brown & Company, Boston & Toronto, 1959, p. 36. For more on key plants, history and humankind, see Bill Laws, *Fifty Plants that Changed the Course of History*, David & Charles, Cincinnati, OH, 2010.
34. Richard Brookes (1730), as cited by Arthur W. Knapp, *Cocoa and Chocolate: Their History from Plantation to Consumer*, Chapman and Hall, London, 1920, p. 165.
35. Chantal Coady, *The Chocolate Companion: A Connoisseur's Guide to the World's Finest Chocolates*, Simon & Schuster, New York, 1995, p. 15. For an overview of these Quaker families and their respective chocolate businesses, see Gillian Wagner, *The Chocolate Conscience*, Chatto and Windus, London, 1987.
36. A little later, E. Harnack, "Zur streitfrage über den fettgehalt in den handelssorten des kakaos", *Deutsche Medizinische Wochenschrift*, 1906, **32**, pp. 1041–1043, makes reference to cocoa as a "curative food". A company known as Sir Hans Sloane Chocolates Limited operates today from Byfleet, Surrey, England. Though their fancy *couverture* chocolates taste considerably different that the drink Sir Hans introduced, they keep the affiliation between Sloane and chocolate alive and well. Their products were featured at several of the London based "Sir Hans Sloane: Discovery, Travels and Chocolate" programme events in autumn 2010 celebrating the 350th anniversary of Sloane's birth.
37. Debauve & Gallais continues to enjoy a Paris-based production of some of the world's finest chocolates.
38. G. B. Wood and F. Bache, eds., *Dispensatory of the United States*, Gregg and Elliot, Philadelphia, 1834, as cited by Martha Makra Graziano, "Food of the Gods as mortals' medicine: The uses of chocolate and cacao products", *Pharmacy in History*, 1998, **40**, pp. 131–146, p. 137.

CHAPTER 5

Chocolate and Nutritional Health: Industrial Era through WWII

> Chocolate is a perfect food .... It agrees with dry temperaments and convalescents; with mothers who nurse their children; with those whose occupations oblige them to undergo severe mental strains; with public speakers, and with all those who give to work a portion of their time needed for sleep. It soothes both stomach and brain, and for this reason, ... it is the best friend of those engaged in literary pursuits.
>
> Baron Hermann von Liebig, Proprietor of Baron Liebig's Cocoa and Chocolate Works, Ltd., London, 19th-Century Promotional Leaflet

> The cacao bean is a phenomenon, for nowhere else has nature concentrated such wealth of valuable nourishment in so small a space.
>
> Alexander von Humboldt, Natural Philosopher and Explorer Extraordinaire

In Mesoamerica, cacao was principally promoted as the "Divine Drink" that provided sustenance and energy. Physician Juanes de Barrios reported that the ancients of Mexico deemed that chocolate was "all that was necessary for breakfast because after eating chocolate, one needed no further meat, bread or drink".[1] But

Chocolate as Medicine: A Quest over the Centuries
Philip K. Wilson and W. Jeffrey Hurst

Published by the Royal Society of Chemistry, www.rsc.org

it was the presumed *medical* benefits of this special substance that prompted its spread across Europe and into North America. Then, beginning in the 18th century and continuing throughout the 20th century, chocolate regained its prominent reputation as a restorative and energy-building drink. J.F.C. Morand promoted chocolate's use on expeditions to provide staying power in his *An Sensibus Chocolate Potus?* (1749). Among the many references to "Health Chocolate" in his influential *Tratado de los Usos, Abusos, Propiendades y Virtudes del Tabaco, Café, Te y Chocolate* (1796), Antonio Lavedan, Royal Surgeon and Director of the Academy of Surgery of Valladolid, Yucatán, proclaimed chocolate as "a food that repairs and fortifies quickly". It was "possible for chocolate alone" to keep humans "robust and healthy for many years", Lavedan claimed, and even "without help from other food, chocolate can prolong life through the great nutrients that it supplies to the body".[2]

5.1 DRINKING CHOCOLATE

Nearly a century and a half later, similar claims were still being made about chocolate. Cadbury's, for example, advertised that chocolate "gives you long staying power" and even "makes strong men stronger". J.F. Beale, Jr, Advertising Director for Pennsylvania's H.O. Wilbur and Sons Chocolates claimed it "has been demonstrated by test" that a good cocoa drink is "cell building brain food, no less than a healthy building, muscle-making, invigorating tonic".[3] Hershey's promoted their chocolate products in the 1930s primarily as "more sustaining than meat" (Figure 5.1). To add substance to these ads, Hershey's published a work in 1925 prepared by the American Food Journal Institute, specifically compiled by Edith C. Williams, that included 166 annotated references to the then current scientific research underlying claims of chocolate's nutritive value as *A Bibliography of the Nutritive Value of Chocolate and Cocoa*. In summary, Williams concluded that chocolate and cocoa were "foods which have gained a rightful place in the diet" (Figure 5.2).[4]

Still, chocolate's potential health benefits were not neglected.[5] Alfred Franklin promoted chocolate's ability to preserve and control health in his 1893 *Le Café, le Thé et le Chocolat*, as did

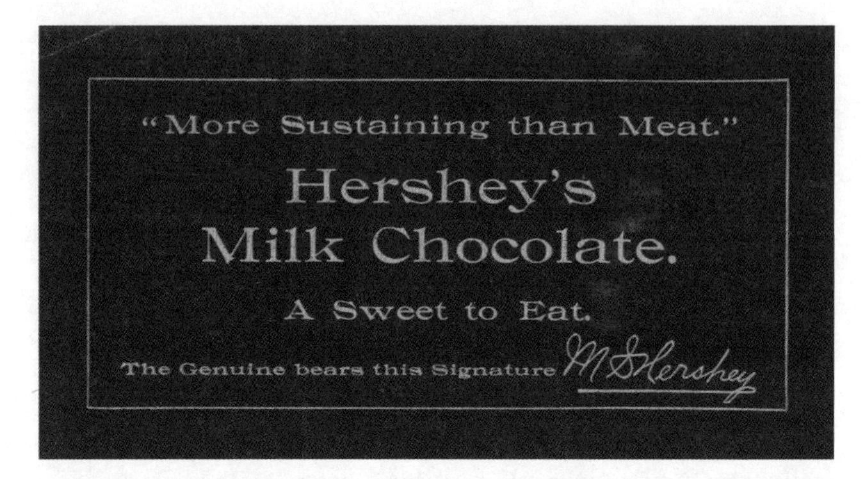

Figure 5.1 Hershey's Chocolate - "More Sustaining than Meat", Hershey Milk Chocolate Brochure (c. 1905).
(Courtesy of Hershey Community Archives, Hershey, Pennsylvania, USA).

Erwin Franke in his 1914 *Kakao, Tee und Gewurze*. At times, chocolate prescriptions and recipes appeared in the same work, such as Thomas Cooper's 1824 *Treatise of Domestic Medicine, to which is added, A Practical System of Domestic Cookery*. Beginning in the late 19th century, a new source of evidence for these benefits became increasingly used – advertising, or what in more recent eras has come to be called direct-to-consumer marketing. These European as well as U.S. advertisements reveal another important conceptual change in chocolate and health, for throughout the nineteenth century they greatly helped solidify what was to become chocolate's enduring reputation as both a medicine and a food. In this way, entrepreneurial advertisers crossed multiple markets. Chocolate became the medicine handed out by confectioners and the food prescribed by physicians. Throughout this chapter the dual recognition of chocolate as both food and medicine is examined. To better distinguish a new form of the evidentiary basis of chocolate's benefits, information for this chapter is drawn primarily from the advertisements themselves. Though individual references are not routinely cited, quotes have typically been taken from newspaper and periodical advertisements as well as from chocolate product containers and wrappers.

Figure 5.2 Cover of Edith C Williams, *Chocolate, Nutrition and Health Bibliography* (1925).
(Courtesy of Hershey Community Archives, Hershey, Pennsylvania, USA).

A century after Dr James Baker established his Dorchester, Massachusetts chocolate company in 1764, "Baker's Chocolate" began to advertise its benefits in providing "an excellent diet for

Figure 5.3 19th-Century Baker's Chocolate Ad Noting the Product to be "Free from the Exciting Qualities of Coffee and Tea".

children, invalids, and persons in health" and in "allay[ing], rather than induc[ing] the nervous excitement" that was a regular consequence "upon the use of tea or coffee" (Figure 5.3). To further convince consumers, manufacturers noted that chocolate came highly "recommended by the most eminent physicians". Commercial agriculturalist, P.L. Symonds deemed, in his *Commercial Products of the Vegetable Kingdom* (1854) – a work written for the Colonist, Manufacturer, Merchant and Consumer – that chocolate is the "cheapest food ... [of which] we can conceive, as it may be termed literally meat and drink". If, as Symonds concludes, our "half-starved artisans and over-worked factory children" were induced to drink chocolate "instead of the innutritious beverage ... tea, its nutritive qualities would soon develop themselves" measurably in those individuals' "improved looks and more robust condition".[6] Physicians, "especially those who, holding office under the Poor Laws", were deemed ideal to "have such large and extensive opportunities for testing [chocolate's] value" over other beverages.[7]

Campaigns emerged to highlight chocolate's benefits as surpassing those of the two other most popular drinks – coffee and tea.[8] Magistrate and gastronome Jean Anthelme Brillat-Savarin – ever the recommender that people "experiment" with their food and drink – specifically noted chocolate's preference over coffee in terms of its nutritional and health benefits. It has "been shown as proof positive that carefully prepared chocolate ... does not cause the same harmful effects to feminine beauty which are blamed on coffee, but is on the contrary a remedy for them". During times of war when chocolate became scarce and expensive, people busied themselves in finding a substitute for this nourishing health drink. But, as Brillat-Savarin concluded, "one of the blessings of peace has been to rid us of the various brews which we were forced to taste out of politeness, and which had no more to do with chocolate than chicory has to do with real mocha coffee".[9] An 1849 *Scientific American* report postulated that those who habitually drank chocolate did not experience attacks of cholera or dysentery as did those who drank only coffee or tea or water. England's *Medical Times* of 28 June 1864 noted that chocolate's "high percentage of nitrogenous matter and its very considerable amount of fat, separate it sharply from tea and coffee".[10] J.B.A. Chevallier concluded his 1871 *Hygiène Alimentaire. Mémoire sur le Chocolat* arguing that chocolate was a complete food, whereas coffee and tea were not. In a paper read before the Surgical Society of Ireland in 1877, Mr Faussett promoted chocolate's ability to take the place of mother's milk, for cacao, in its "natural state" was claimed to be "abound[ing] in a number of valuable nutritious principles, in fact, in every material necessary for the growth, development, and sustenance of the [child's] body" (Figure 5.4).[11] Dove's Chocolate, manufactured by the Blumenthal Brothers of Philadelphia, similarly touted that this "perfectly digestible" product was the "best substitute for tea or coffee" being, in general, "more healthy", and that it was specifically "recommended by physicians as medical nourishment for invalids". Wilbur's Chocolate Company of Lititz, Pennsylvania also set chocolate apart from the other popular drinks, noting that their chocolate was "recommended by prominent physicians throughout the country as an anti-dispeptic substitute for coffee and tea". In the early 20th century, the nearby Hershey's Company was also promoting their product as "an excellent substitute for tea and coffee".[12]

Figure 5.4 Late 19th-Century Nestlé's Chocolate Ad – a "Food for Infants".

Elevating chocolate's powers over that of the two other popular beverages was also an aim of Fannie Farmer's popular *The Boston Cooking School Cook Book* (1896), in which she argued that "cocoa and chocolate differ from tea and coffee inasmuch as they contain nutriment as well as stimulant". She spoke of many individuals who "abstain from the use of tea and coffee [but who] find cocoa indispensible". Farmer continued, claiming that not only is chocolate "valuable for its own nutriment, but for the large amount of milk added to it. Cocoa may well be placed in the diet ... of a child after his third year", she claimed. Moreover, "invalids and those of weak digestion can take cocoa [with milk] where [pure] chocolate would prove too rich". Individual product ads of the 1890s made similar claims, urging consumers, "Don't wreck your nerves with tea and coffee, or ruin your stomach with trashing substances", when you can consume cocoa, a "natural food drink" which "makes rich blood and strong nerves".[13]

5.2 EATING CHOCOLATE

One notable change in the form of chocolate consumption resulted when this product was no longer available simply as a drink but could also be ingested in the form of a candy bar. Previously, people had commonly drunk chocolate as both a "confection" as well as part of a mixture used to "coax ... [medicine] into palatability" after having added a few pleasant tasting spices and, in particular, sugar, to the mix.[14] At least as early as 1826, edible chocolate became available in England in the form of lozenges that were deemed "a pleasant and nutritious substitute for food while traveling or when unusual fasting is caused by an irregular period of mealtimes".[15] Suchard's of Neuchâtel, Switzerland gained early renown for their sensationally popular chocolate lozenges, "diablontins". Dr Joseph Fry of Bristol, England, had been producing drinking chocolate since 1728. Beginning in 1795, his company, Fry & Sons, was one of the first to use the Industrial Revolutionary James Watt's steam engine (fully developed in 1775) to power thc grinding of cacao beans. In 1845, Fry & Sons introduced the first such concoction known as "eating chocolate", advertising these edible bars as "chocolat dèlicieux à manger".[16]

The Swiss inventor, Daniel Peter reached the end of an eight-year series of experiments in 1875 designed to establish a perfect combination of milk and chocolate. His experimental product, solid milk chocolate, which was a mixture of Henri Nestlè's newly prepared condensed milk, cocoa, cocoa butter and sugar became quite popular. Challenging coffee's prior success, Peter's chocolate was "now a regular part of people's diet, but its nutritional value is higher [than that provided by coffee alone.]" Conveying a strong sense of Swiss pride, Peter's chocolate ads were designed to persuade consumers that this "Original" milk chocolate was "a delicacy and a food in one luscious combination" that was "as distinct from ordinary eating chocolate as the Alps are from foot-hills". Thanks primarily to Peter's work, the Swiss held a monopoly on "closely guarded secret formulas" of milk chocolate until England's Cadbury's Dairy Milk Chocolate rapidly gained a strong following after its introduction in 1904. Cadbury's – the milk chocolate that "makes strong men stronger" – was marketed as "the most refreshing, nutritious and sustaining" of all milk chocolates.

Chocolate's duplicity as medicine and food held sway for years within the advertising market. The nineteenth-century Cocorico bar was advertised as "recommande aux enfants & aus malades" as well as "constitute un ailment hygiénique par excellence". Like other contemporary chocolate products, it was recommended as "délicieux pour malades & bien portants". A 1875 chocolate "Milk Food" label reported, "Farine Lactèe Nestlè ailment complet pour les enfants en bas age". Rothwell's Milk Chocolate was advertised as "a sweetmeat and a food". Hershey's promoted its early 20th century chocolates as "A Sweet to Eat, A Food to Drink". Cadbury ads urged their Dairy Milk Chocolate consumers to "Eat More Milk". Given our present fixation on chocolate's potential cardiovascular benefits, it seems that some early 20th century ads presciently, though unknowingly, conveyed something of a double meaning, as in the cupid-clad ad for Versailles chocolate, Pascal, which meets "The heart's desire". Johnston's Chocolate ads of 1925 also recommend their R.S.V.P. brand for all "Affairs of the Heart".

5.3 PURE AND UNADULTERATED CHOCOLATE

The 19th century experienced a boom in improved methods for manufacturing larger quantities of chocolate, all the while manufacturers were fraught with anxiety over maintaining the quality of their products (Figure 5.5). Concurrently, chocolate's medicinal

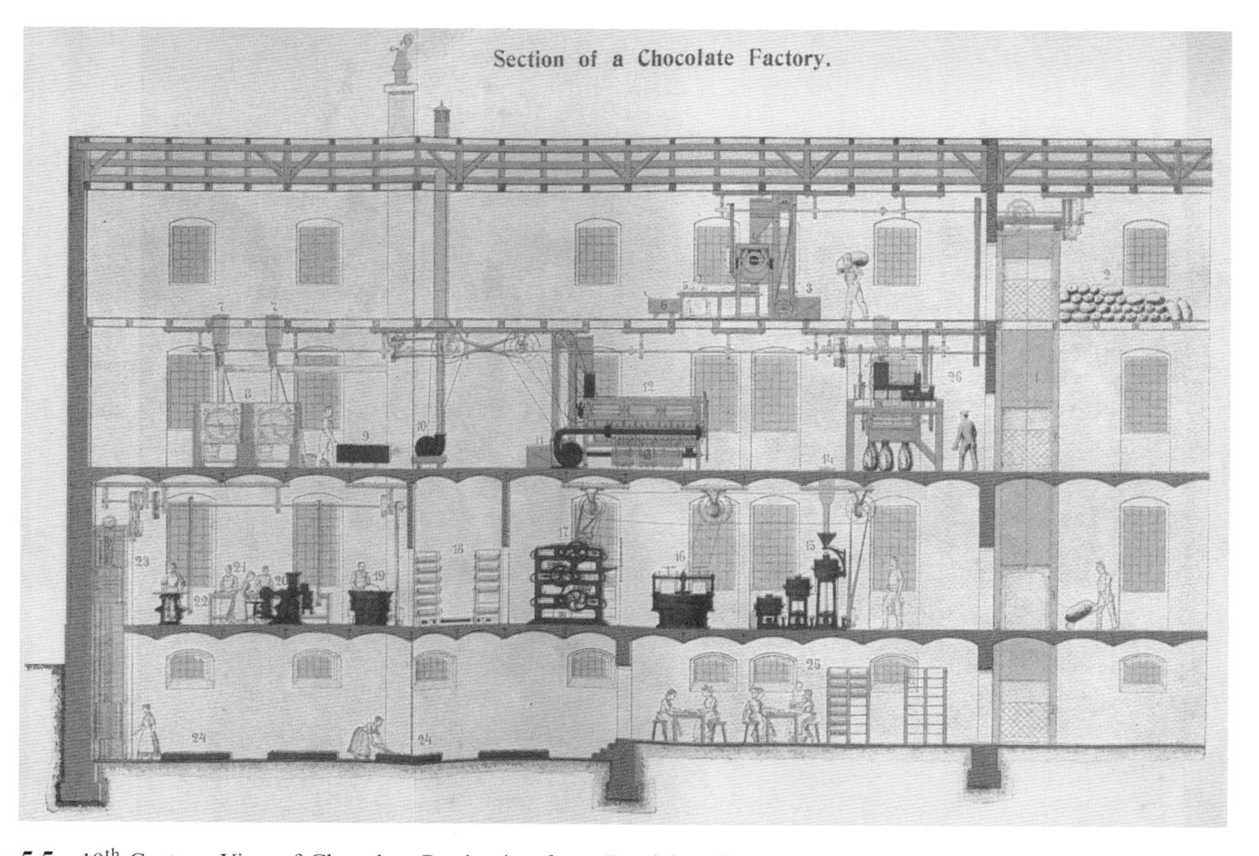

Figure 5.5 19th-Century View of Chocolate Production from Receiving Cacao Beans on Top Floor to Product Packaging on the Ground Floor. Messrs Lehman Factory of Dresden, included in Dr Paul Zipperer's *The Manufacture of Chocolate* (1915). (Courtesy of Hershey Community Archives, Hershey, Pennsylvania, USA).

benefits in terms of "pure" ingredients became increasingly advertised in promotional literature. As such, these ads represent a shifting emphasis in the evidence used to support chocolate's healthiness.

In an era of antigerm campaigns of the early 1900s, one newly formed industrialist company in the United States began to promote the health and medical benefits of chocolate. Milton S. Hershey represented his milk chocolate drinking cocoa and his "Great American Chocolate Bar" as pure, natural, nutritional and wholesome. He advertised the nutritious qualities of chocolate against a background of green fields, cows and wholesome country milk. Milk, fortunately, was white – the color that the medical community had selected to represent clean, sterile and sanitary environments as part of their public health campaigns (Figure 5.6).

> The beautiful cows in the pasture fed,
> Clean as could be from their tails to their head.
> Making pure milk early and late
> For making Hershey Cocoa and Chocolate.

A further verse in this advertisement conveyed a similar idea:

> A child said to her mother dear,
> "Cocoa and Chocolate are healthful, I hear".
> "Yes", said her mother, "if they're pure and fresh".
> Said the child, "Well, that means Hershey's I guess".

A promotional booklet for Hershey products publically displayed in a recent "The Kiss Story" exhibit at the Hershey Museum also focused on the healthy benefits of chocolate in an entry titled, "A Word to Mothers" (Figure 5.7).

Milton S. Hershey's drive to prepare healthy products was not new. Since the 1880s, Hershey had been producing and advertising a wide variety of "medicated candies" in the form of cough drops. According to a 1881 catalogue, Hershey manufactured "cough confection" products for sale at his Beach Street, Philadelphia premises including: Horehound Drops, Flaxseed Drops, White Oak Bark, Silver Tar Drops, Colt's Foot Rock, H[enry] H[ershey] Drops, Tar Beans, Horehound Beans, Boneset Beans, and Rock and Rye Cough Drops.

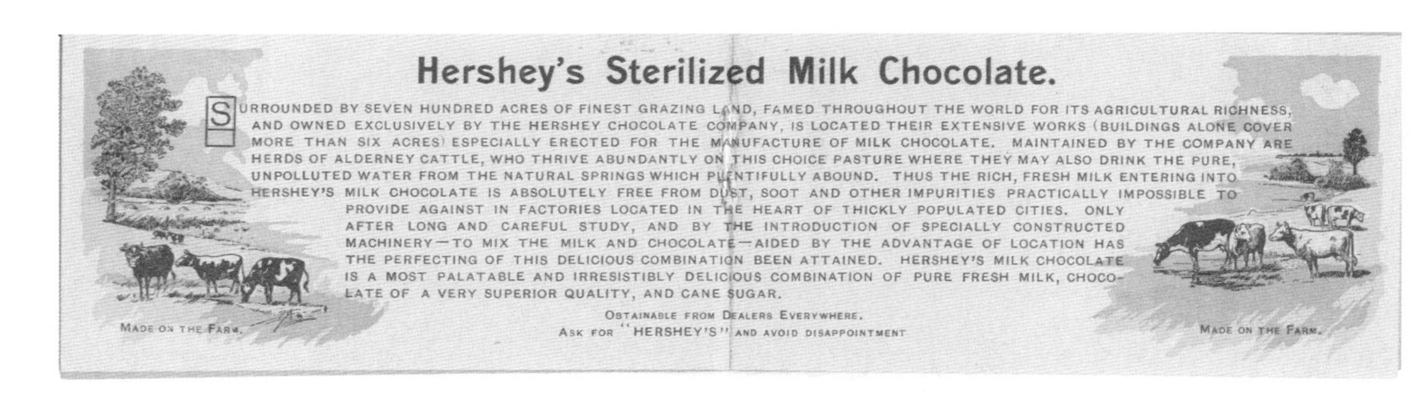

Figure 5.6 Milk and Farm Fresh Countryside, Hershey Milk Chocolate Brochure (c. 1905). (Courtesy of Hershey Community Archives, Hershey, Pennsylvania, USA).

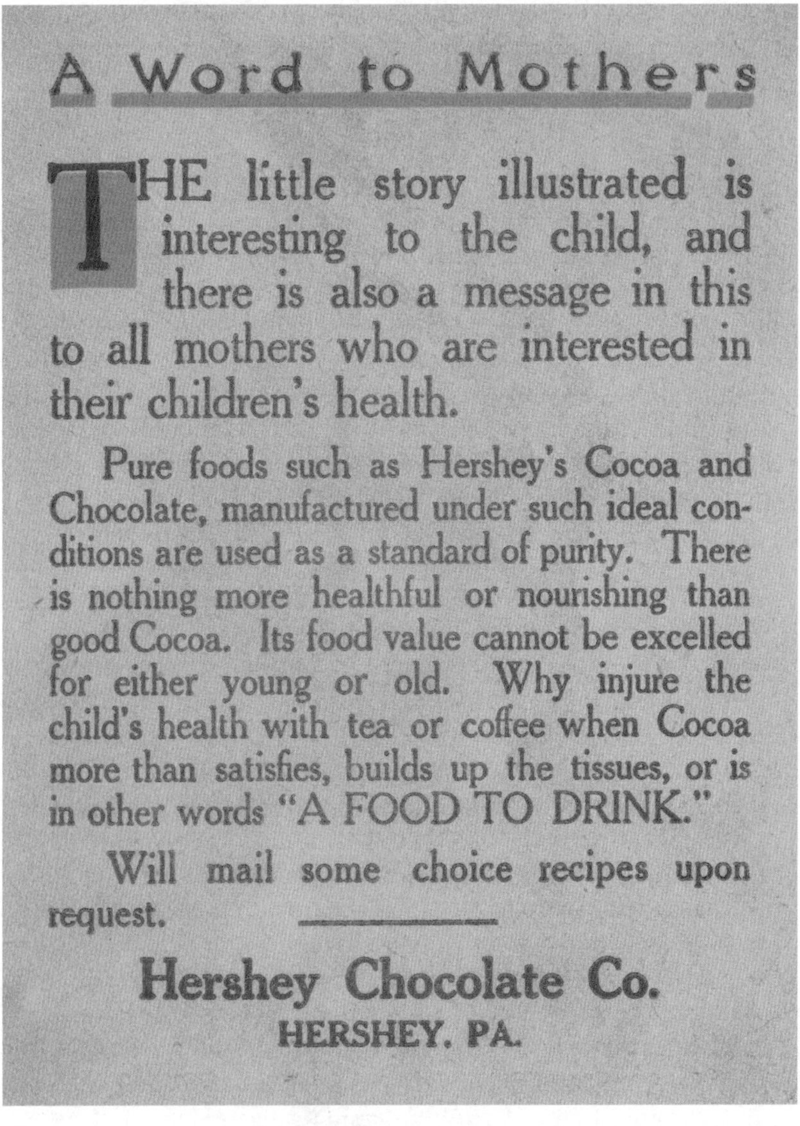

A Word to Mothers

THE little story illustrated is interesting to the child, and there is also a message in this to all mothers who are interested in their children's health.

Pure foods such as Hershey's Cocoa and Chocolate, manufactured under such ideal conditions are used as a standard of purity. There is nothing more healthful or nourishing than good Cocoa. Its food value cannot be excelled for either young or old. Why injure the child's health with tea or coffee when Cocoa more than satisfies, builds up the tissues, or is in other words "A FOOD TO DRINK."

Will mail some choice recipes upon request.

Hershey Chocolate Co.
HERSHEY, PA.

Figure 5.7 "A Word to Mothers" from Hershey's "Green Grass Jingle Book" Promotional Booklet (c. 1915).
(Courtesy of Hershey Community Archives, Hershey, Pennsylvania, USA).

Yet, long before Hershey's entry to the world of chocolate, the purity of chocolate products had gained considerable attention. During the reign of England's George III, an act was passed legislating that "if any article made to resemble cocoa shall be found in the possession of any dealer, under the name of "American cocoa" or "English cocoa", or any other name of cocoa, it shall be forfeited, and the dealer shall forfeit £100".[17] Gastronome Alexandre-Balthazar-Laurent Grimod de la Reynière warned, in his *Almanach des Gourmands* of 1805, that though chocolate was pure when trusted to the hands of apothecaries, much of it sold elsewhere was tainted with impurities.

Concerns about the purity of chocolate had appeared sporadically in the medical literature since the time of William Hughes' *The American Physitian [sic]* (1671). There, Hughes had warned that "the addition of many ingredients" into concoctions of chocolate drink did not in "any way at all advantage the wholesomeness" of the "chiefest ingredient, the Cacao". By adding various ingredients to the cacao, Hughes noted, the "property thereof from what it naturally is in it selfe is quite changed". If physicians would but "narrowly pry into the secrets of the nature of" pure cacao, they would resist adding further ingredients. Indeed, it was the "adulteration of this [American] Nectar which undeservedly makes it ill thought of", so Hughes exclaimed.[18]

In the mid-1800s, anxiety grew over the unhealthy chemicals and fillers that, when mixed during the production of cocoa powder, overcame chocolate's natural medical benefits. In 1851, the English medical journal *Lancet* reported the results of an analysis of 50 commercial chocolate products.[19] The findings suggested that 90% of these products were adulterated during their production with a variety of fillers including "starch, animal fat, red and yellow ochre, red lead, vermillion, sulphate of lime, and chalk".[20] Others reported on finding fillers including brick dust (for coloring), coconut oil, clarified mutton fat, suet and tallow – the latter fillers used as replacement for the cocoa butter that had been removed during the production of cocoa powder. It is ironic that chocolates prepared using the most adulterated substances – like Du Barry's Revalenta Arabica – were advertised with "the most extravagant [health] claims". In this case, the London East End producers claimed that Du Barry's Revalenta Arabica "purifies and improves the blood, strengthens the stomach, nerves and muscles, removes all gastric or nervous irritability, ensures tranquil slumbers, absorbing and eliminating all acidity, feverishness, headaches, lassitude, constipation, dyspepsia, sleeplessness and low spirits".[21]

A.H. Hassall drew further attention to this concern in his 1855 *Food and Its Adulteration*. Manufacturers were not explicitly attempting to diminish chocolate's benefits by using additives, Hassall argued, but rather, many were explicitly attempting to reduce the high taxes that their countries had to pay for cacao by substituting some of the cacao in their products with cheaper, locally available filler materials (Figure 5.8). In 1891, *Peterson's Magazine* propagated these anxieties to an even larger reading

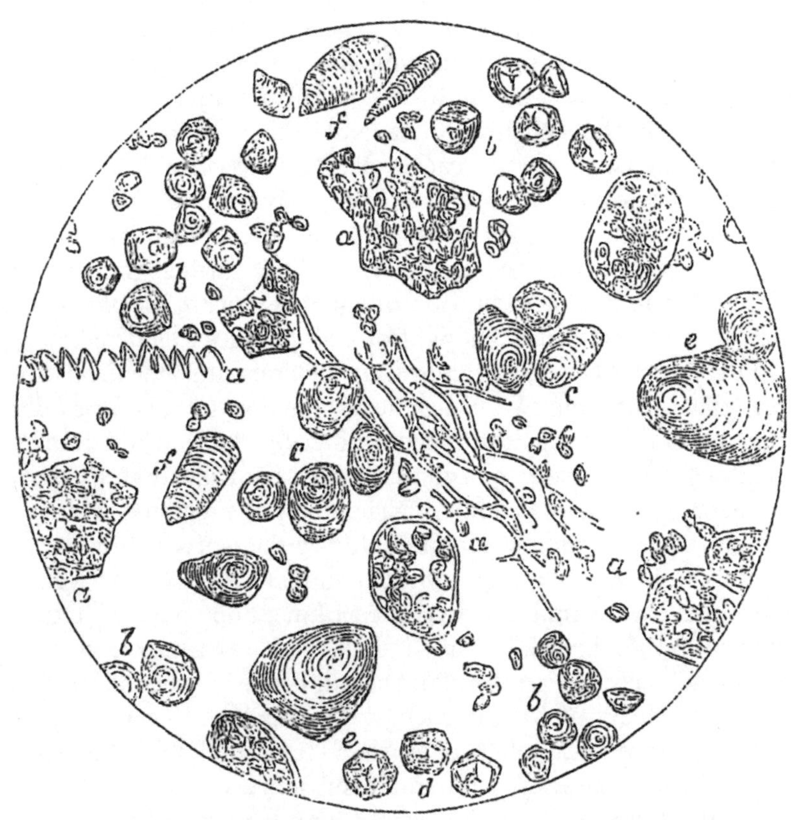

DUNN'S GENUINE UNADULTERATED CHOCOLATE POWDER.
a a a, starch-granules and cells of cocoa; *b b b*, granules of *tapioca-starch*; *c c*, *Maranta arrow-root*; *d*, *Indian corn meal*; *e e*, *potato-starch*; *f f*, *Curcuma arrow-root*.

Figure 5.8 Microscopic View of "Dunn's Genuine Unadulterated Chocolate Powder", from Arthur Hill Hassall's *Food and its Adulterations* (1855).

audience. There, readers found distinctions between two categories of chocolate adulteration:

> 1) Those that were "simply fraudulent, but not necessarily injurious to health" by using "some cheap but wholesome ingredient [mixed] with the pure article for the purpose of underselling and increasing profits"; and
>
> 2) Those that were "injurious to health", by using "drugs or chemicals for the purpose of changing the appearance or character of the pure article, as for instance, the admixture of potash, ammonia, and acids with cocoa to give the apparent smoothness and strength to imperfect and inferior preparations".[22]

As is often the case in reform efforts, progress, at least in terms of better assuring the healthiness or purity of products, moved with glacial swiftness. W. Clarke Saunders was still proclaiming in 1895 that the "monstrous fraud perpetuated upon the poor by the adulteration of cocoa cannot be over-estimated".[23] In the United States, it was not until 1906 that the Pure Food and Drug Act (based on the Heyburn Pure Food Bill) was passed – an Act that was frequently re-examined and amended five times before being superseded by the Pure Food, Drug and Cosmetic Act in 1938. In the 1906 Act, a food and drug – like chocolate – was considered to be adulterated if

> mixed or packed with another substance to lower its quality; if any substance was substituted by another; if any valuable constituent was wholly or in part omitted; or if it was mixed, colored, coated, or covered in any way to conceal inferiority or damage.[24]

Following these federal acts, marketing the healthiness of chocolate products in terms of "purity" rapidly became the norm. In 1908, the Sun Shine Biscuit company introduced chocolate-containing Hydrox Cookies, advancing the blended names of hydrogen and oxygen as trusted symbols of the purity of their product. Hershey's became the "first chocolate and confectionary company to voluntarily provide nutritional labeling on food labels".[25] A 1920 Hershey's Cocoa ad read, "When we analyze Pure Cocoa we find it a perfectly natural food. Pure Cocoa is easily digested. Be sure that you get it Pure. Avoid Cocoas containing hops, kola and like substitutes". And consumers, rest assured, "Hershey's Cocoa is Pure" (Figures 5.9 and 5.10). The use of their "Baby in the Bean" trademark was also meant to "enable the discriminative purchaser to become an expert in the detection of the genuine from the imitation". Hershey's further assured consumers that their products were free of those "alkalis or chemicals – which by some manufacturers are used as an economical method of eliminating the indigestible substances", yet "have a tendency to destroy much of the flavor inherent to good cocoa". Similarly, Wilbur's cocoa was, according to a 1919 ad, "pure and nourishing, and entirely free from harmful stimulants or substitutes of any kind". Though theobromine in cacao was a stimulant, presumably Wilbur's did not perceive it as a "harmful" stimulant.[26] A later Wilbur's product informed consumers of their attention to purity, openly attesting that their drink mix "complies with all requirements – including a minimum of 22% cocoa butter, of the Food, Drug & Cosmetic Act, effective June 25, 1939, as amended". Liggett's of New York City manufactured their "Delicious Cocoa" sold, so their product label informs us, in "America's Greatest Drug Stores". Furthermore, their label touts that their cocoa was "manufactured by a process which retains all of the palatable and healthful properties" of the cacao bean. For the Rawleigh's Company of Freeport, Illinois, the name of their product, "Good Health Cocoa", simply says it all.

The connotation of the word "purity" in relation to cocoa powder and chocolate initially conveyed various meanings. To strive for a conformity of understanding, standardisations of purity had to be established. Many of the standards that were listed in pharmacopoeias and other comparable resources resulted from consensuses of meaning decided at professional gatherings. One of the first post-FDA Act gatherings was the Congress of Cocoa and Chocolate Makers that gathered in Berne in 1911 and established the standards of purity for their products that held sway for many years. Four years later, in the United States, federal food authorities including agricultural chemists worked as part of a Joint Committee on Definitions and Standards towards adopting a guide for officials in the Department of Agriculture to use as a means of

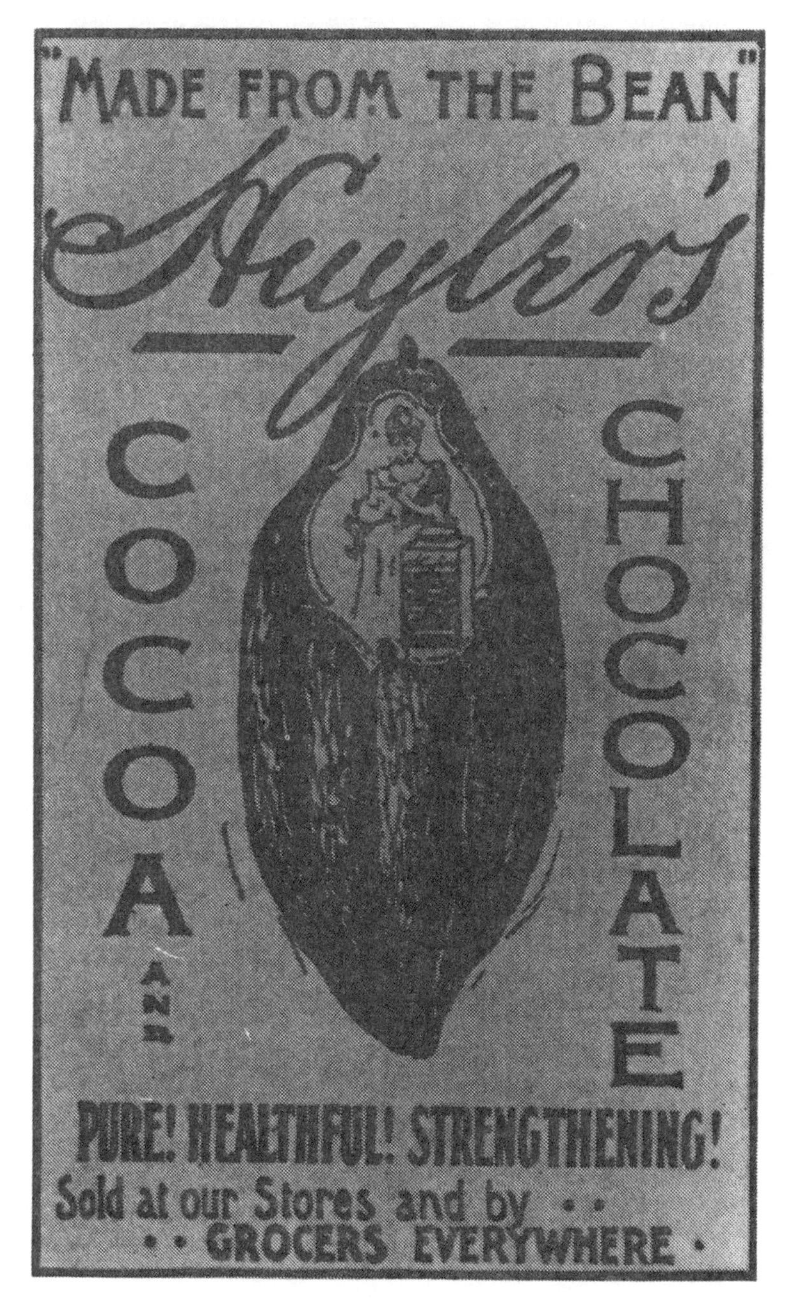

Figure 5.9 Purity is Featured on this 19th-Century Huyler's Chocolates of New York Ad. Milton S. Hershey began working at Huyler and Company in New York in Spring 1883, returning to Lancaster, Pennsylvania in 1885.

Figure 5.10 Hershey's Cocoa is Pure and Unadulterated, Hershey Bar Card (c. 1909–1918).
(Courtesy of Hershey Community Archives, Hershey, Pennsylvania, USA).

enforcing the 1906 Act. Though not specified in these standards, concern about maintaining freshness of cacao beans and of chocolate products also remained of critical importance (Figure 5.11).

It was also at this time that initial advances in chemical technology allowed a better analytical determination of the composition of cacao itself. J.S. Bainbridge and S.H. Davies initially identified 12 chemical compounds within the bean.[27] Throughout the century

Figure 5.11 National Confectioners Association Laws Fighting Against Adulterations (1888).

an explosive growth of knowledge, relative in large part to new technologies, occurred regarding the chemical composition of cacao such that we now know of over 500 compounds within the bean. (See Appendix 3.)

5.4 WHOLESOME AND NUTRITIONAL CHOCOLATE

Chocolate's healthiness in terms of its nutritional value became paramount to early and mid-20th century advertisements. Some, like this 1918 Cowan's Cocoa ad, addressed more general wholesome and healthy nutritional values.

> Life's greatest assets are Health and Strength and without these, existence becomes intolerable.
>
> The human body, under the best of conditions, is a fragile structure, easily susceptible to climatic conditions, over-heating, exertion, [and] mental and physical emotions. This subject requires constant attention if health and strength are to be continually maintained. The most sensible method of preserving health is to consume food which produces it. The food which produces health is that which contains Carbo-Hydrates, Proteins, and Fats. Cocoa is a palatable liquid food containing, when mixed with milk, all these necessary substances in a form that is not injurious to the weakest digestion. For the easiest and most pleasurable way to obtain just the right kind of nourishment the body needs, drink Cocoa. For the best and quickest results drink Cowan's Perfection Cocoa.

Runkel Brothers of New York City assured consumers that in preparing their Breakfast Cocoa, the "excess oil (cocoa butter) removed" in processing "does not lessen the quantity of nutritious substances" or modify the "Theo-bromine content in the Cocoa Beans" that "constitute its nutritive qualities". Rather, removing the components as they did, made their cocoa more "easily digested" and thus more capable of providing a more "perfect food which is readily assimilated by the system of persons in health or of invalids". They concluded with the rhetorical claim that because of this special processing, "all physicians recommend Runkel Brothers Cocoa".

At times, specific physician's recommendations were included on the product ads, thereby attributing testimonial support to figures of authority that should simply be trusted. The Curtiss Candy Company of Chicago promoted one of their major products, the Baby Ruth chocolate candy bar (which appeared in 1920 and was named after President Grover Cleveland's daughter, Ruth – not as urban legend would have us believe, after the American baseball great, Babe Ruth) with the authoritative voice of Dr Allan Roy Dafoe. Dr Dafoe championed this product as being nutritionally "rich in Dextrose, [a] vital food-energy sugar, and other palatable ingredients" all of which combined to make this a "pleasant, wholesome candy for children". Adding further rhetorical support was the fact that Dr Dafoe was physician to the world-wide sensations, the Dionne Quintuplets who were fittingly featured in the ad as well (Figure 5.12).

Another Chicago-based candy confectioner, the Cook Candy Company, offered the nutritious "Vita-Sert" bar in the 1940s. This "double-feature" chocolate aimed to satisfy "both your cravings for sweets and your need for vitamins". The "Rich, energizing

Figure 5.12 Dr Allan Roy Dafoe promoting Curtiss Candy's Baby Ruth (c. 1941).

chocolate" was mixed with a "full day's supply of vitamins". This wrapper information was further enhanced in news ads which specified that each chocolate bar provided "100%" of the "minimum adult daily requirements" of six "essential vitamins" (A, B-1, B-2, C, D and "Niacinamide") as "set by the U.S. Government".

Mid-century newspaper and magazine ads delved even further into chocolate's nutritive value. Edith C. Williams, Director of the American Food Journal Institute was quoted in ads claiming that chocolate had "gained a rightful place in the diet". When combined with milk, these products offered "a particularly nourishing food which builds bone, gives energy and builds tissue". Children who naturally "need more protein and less energy than the grown-ups" simply need chocolate milk, for "the milk contains valuable minerals for bone-building", and the chocolate helps to make milk "an attractive beverage".[28] Elsewhere, Caroline B. King, a "leading American dietician" added comments supporting the view that chocolate "contains qualities which at once recommend it to the dietician".[29] To even further convince consumers, one mid-century Hershey's ad specified chocolate's food values in comparative tables (Figure 5.13).

Given the rise of milk chocolate products, it was not just the purity of chocolate but also that of sugar and milk that came under scrutiny. In order to "secure the best and most digestible chocolate", so Hershey's argued, it was "necessary to have cane sugar of superior grade". In order to ensure the quality of sugar used, this company purchased 10 000 acres of sugar plantations in Cuba in 1916. Cuba, the company argued, was "by nature best fitted for the production of sugar".[30] As to the milk used in their products, "all the cows that furnish milk for the [Hershey] factory ... are subject[ed] to frequent inspection so that by no possible chance can any milk be delivered that is not up to the high standard of purity" that their "sweet milk chocolate demands" (Figure 5.14).[31]

5.5 HOMEOPATHIC AND DIETETIC CHOCOLATE

One other change regarding the mandatory reporting of ingredients in chocolate products affected the labeling of a long-standing product, Homeopathic Chocolate. In the early 1800s, Homeopathic medicine began to challenge the prevailing drug therapies. German physician Samuel Hahnemann conducted

TABLE OF FOOD VALUES

			CALORIES
1	Pound Hershey's	Milk Chocolate,	2335
1	" "	Sweet Coating,	2165
1	" "	Almond Chocolate,	2600
1	" "	Breakfast Cocoa,	1890
1	Cup "	Breakfast Cocoa,	185
1	Cup Coffee, with Cream and Sugar,		50
1	Cup Tea,		Little, if any
1	Pound Lean Beef,		1105
1	" White Fish,		475
1	" Oysters,		235
1	" Beets,		160
1	" Potatoes,		295
1	" White Bread,		1200
1	" Apples,		290
1	Dozen Eggs,		1180

A Calorie is the unit measure of food value.

HERSHEY CHOCOLATE COMPANY
HERSHEY, PA., U. S. A.

Figure 5.13 Hershey's "Table of Food Values", from Macfadden's "Hershey, The Chocolate Town" (c. 1923).
(Courtesy of Hershey Community Archives, Hershey, Pennsylvania, USA).

Figure 5.14 Sterilisation and Sanitary Milk Production, Hershey Bar Card (c.1914–1918).
(Courtesy of Hershey Community Archives, Hershey, Pennsylvania, USA).

experiments on himself and his family to determine the minimal concentrations of medicines needed to effectively cure disease. He viewed disease as "a disruption in the body's ability to cure itself", and he sought to administer the least possible concentration of a drug necessary to promote this self-healing.

Determining the precise dosages required to cure ailments came through a method that Hahnemann and later Homeopaths called "proving". Here, through experimental trial and error, Homeopaths noted how by administering a drug to the state of its becoming poisoning to the body, they saw (or "proved") that the effects of the poisoning were similar to the symptoms of the disease that it cures. By conducting provings on himself, his family, his friends and later, his students, Hahnemann differentiated which precise remedy would best fit each individual case history.

After its introduction in the 1820s, many physicians converted to Homeopathic principles. Medical schools throughout the U.S. began teaching Hahnemann's theory, using his *Organon of Homoeopathic Medicine* (1810) as their primary text. Over time, many of those schools came under criticism by a variety of medical education policing organizations. Consequently, this once common sectarian medical practice ebbed throughout much of the U.S., though homeopathy has retained its popularity in many countries, including Great Britain, the Netherlands and India.

Chocolate that had been tested in terms of provings in the 19th century was commercially distributed as "Homeopathic" chocolates. According to Hassall's survey of British made chocolates, Homeopathic chocolates were being manufactured by a wide range of companies including Leath's, Cadbury's, Graham & Hedley's, J.S. Fry & Sons', Nicol's, Taylor Brothers', Barry & Company, Relfe's, White's, Steane, Davis & Company, and Epps' (Figure 5.15). In essence, it was one of the chocolate brands offered by all leading manufacturers of the day. One producer specified that its Homeopathic chocolate was prepared from the whole seed nib and husk ground together until it gained a flake-like appearance after which arrowroot was added.[32] At times, products were sold as "homeopathic" or "soluble" cocoa after the cocoa butter had been removed. Such labelling may well have expedited marketing during the early Homeopathic boom, but as G.A.R. Wood noted, they were falsehoods for, in fact, "cocoa powder does not dissolve".[33]

"Proving" was a concept that Hahnemann conceptualised as part of the Homeopathic method of remedy analysis and development. Chocolate provings, common in the 19th century, are still carried out today. Jeremy Sherr, Proving Director of England's Dynamis School for Advanced Homoeopathic Studies, published *The Homoeopathic Proving of Chocolate* in 1993. In it, he described the results of a four-year process by which students at that school "proved, supervised, extracted and collated" chocolate – "always striving for the highest possible degree of accuracy and finesse" in their work. In formatting the process, Sherr noted a "major dilemma" he faced in choosing whether to analyse processed chocolate or the cacao bean. It is "desirable to prove purer substances rather than complex ones", yet since it is processed chocolate that is "craved by so many", it is that substance which may "contain other ingredients important to the proving". Thus, he ultimately selected to "prove" processed chocolate.

Typically, provings result in a "pure document, devoid of personal interpretation" from which it is "left to each individual homoeopath to weave the symptoms together into a meaningful picture". Diverging from this pattern, Sherr noted some significant "threads ... running through this [chocolate] proving". Namely, "feelings concerning the texture, smoothness, warmth and 'melt in the mouth' qualities" were the "most venerated", and were commonly "associated with nourishment, breastfeeding and motherly love".

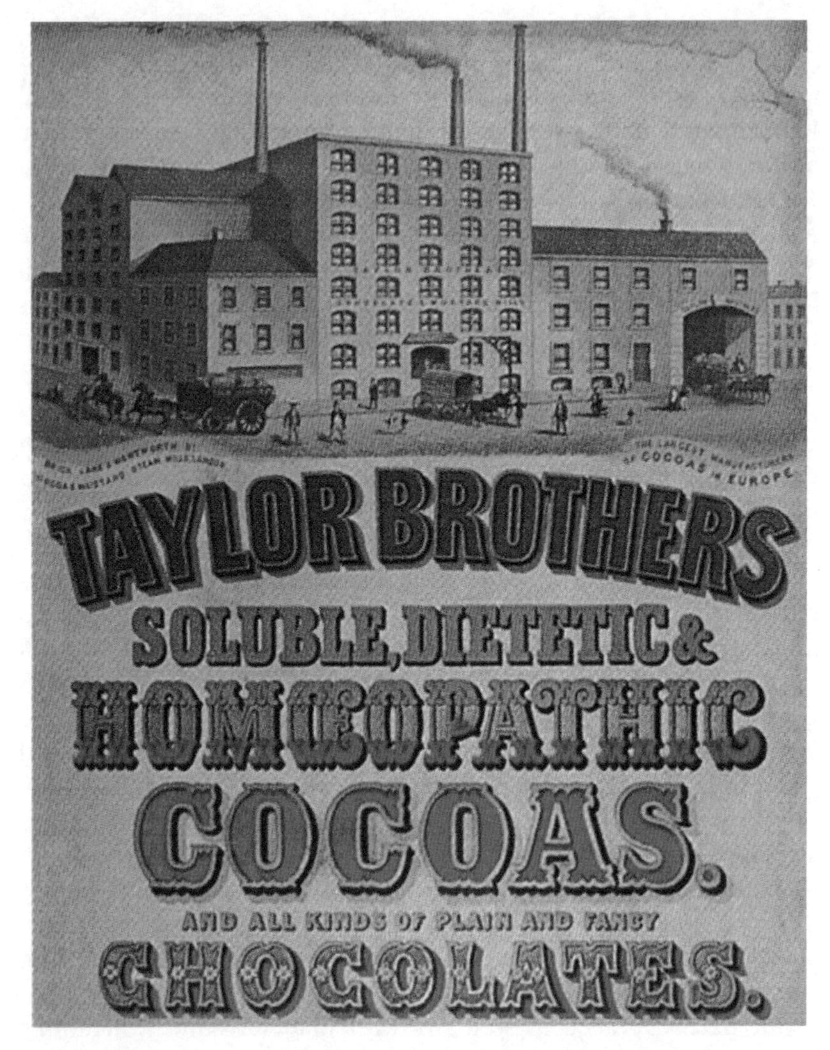

Figure 5.15 Taylor's 19th-Century Homoeopathic Chocolate Trade Card.

Such claims closely resemble reports within concurrent allopathic medical studies which suggest that chocolate's power lies in the combination of "sweet taste tempered by a touch of bitterness" together with a "complex aroma and sensuous mouth-feel".[34] Among the conclusions, Sherr hoped that chocolate which is "now such an integral part of modern civilisation" will soon be shown by all measures to "provide us with another efficient agent for the healing of human kind".[35]

Many of the ads during the 19th century also featured "dietetic" chocolate alongside their Homeopathic products. Unlike what such term most readily brings to mind today, "dietetic" in the 19th century implied the product's relative benefit to general bodily health when added to the regular diet. As economic historian J. Othick described, "moving cocoa out of the medicine cabinet and into the larder" made great economic sense. From the manufacturers' point of view, chocolate consumption was "bound to rise if people were taking the beverage on a regular basis rather than just when they were feeling under the weather".[36]

5.6 POWER, PEP AND CHOCOLATE

Building upon the concept of chocolate being good for the diet, a slowly emerging rhetoric surrounding name brands of chocolate products became detectable during the late 19th century when products such as Neal's "Vigor" Chocolate were promoted. During the 20th century, merchandising chocolate by nutritional or health-related brand names gained considerable momentum. Schrafft's claimed in 1931 that their chocolates provided "one of nature's shortest cuts to stimulation through food". Thus, "for your health's sake, keep a box handy when you work or play". Chocolate was also the perfect "impromptu breakfast" when one was in a hurry or travelling.[37] Later in the century, chocolate breakfast drinks sold in both liquid and powder form became a nutrition fad in many countries.

Similar chocolate products available at this time were claimed to be "packed with energy reserve" and "pep", to "get em going – keep em going" including Ovaltine and the Clark Bar.[38] Some chocolate bars also conveyed messages of power and pep in their very name, such as Three Musketeers (Mars), Power House (Walter H. Johnson Candy Company), the Bolster Bar (D.L. Clark Company) and the Pep Up Candy Bar (Trudeau's). Images of physical fitness in the form of bicyclers, baseball players and other athletes regularly appeared in chocolate ads to help identify specific products with an ability to add a quick and wholesome energy boost. Frederick Cavill, in his celebrated swim across the English Channel in 1876, valued chocolate – Cadbury's chocolate in particular – to be the most concentrated and sustaining food upholding his staying power.[39] Other ads more indirectly suggested their power, such as when Butterfinger was promoted as the chocolate that "accompanied Admiral Richard Byrd to the South Pole in 1928".

This energy-laden rhetoric supporting chocolate's utility was about to take a quick turn. By 1940, Nestlés proffered their chocolate as a "Fighting Food", featuring a comparative table of the energy provided by a lamb chop, milk, eggs, bread and their chocolate. The chocolate bar, they argued, has "come into its own on every fighting front of the war" since it was able to offer "maximum nourishment with minimum bulk". There is "more quick energy packed into the familiar chocolate bar than is contained in many recommended energy foods". As such, it became "one of the answers to the problem of keeping the soldier supplied with food in modern, high-speed warfare". 1943 ads for the Mars bar proclaimed that their "chunks of sheer delicious goodness" were "made with chocolate to sustain, glucose to energize, [and] milk to nourish" the troops. As war diaries and histories inform us, German soldiers – particularly during the early years of the war – also had chocolate bars in their rations.

Hirshfield's "Tootsie Rolls", available since the 1890s, proved to be a "great source of quick energy" that were "favored by soldiers" as they could be "kept in pockets or barracks bags and didn't melt". On a less glamorous side, War Bond era ads also noted that one Tootsie Roll contained "as many energy units as a woman used during nearly two hours of ironing". Other ads noted, "wherever our fighters go, Baby Ruth goes too", adding the disclaimer that, "When you don't find Baby Ruth on the candy counter, remember ... Uncle Sam's needs come first with us and with you". Further to Uncle Sam's needs, Baker's Chocolate Company produced their Flying Fortress Bar (each with the actual image as well as a "Spotter's View" of the Boeing B-17 Bomber) as the "nourishing, sustaining food" that provided an "ideal addition to the kits of Service Men, Air Raid Wardens, and Sportsmen".

WWII was certainly not the first war period during which the healthy benefits of chocolate were noted. One-ounce chocolate-sugar cakes (with equal portions of each named ingredient) were widely distributed as emergency rations during the Great War (WWI). Armies from all nations had incorporated chocolates into rations for food and stimulant, particularly when troops were temporarily removed from their base supplies.[40] At that same time,

Baker's produced chocolate bars stamped "WTW" – Win The War.[41] "Sugar-sweetened slabs of chocolate" had been issued as British seamen's rations as early as 1780, with "half of the cacao imported into Great Britain" being "allotted to the Royal Navy" during the mid-1800s.[42] A. Debay (1864) promoted the addition of cocoa to soldiers' emergency rations.[43] Queen Victoria sent her troops in the South African Boer War one half million pounds of chocolate for Christmas in 1899. During this same era, the U.S. Government sent tons of chocolate to U.S. forces fighting in the Spanish-American War (War of Philippine Independence) in Manila. Curiously, Admiral Dewey – the "Hero of Manila" – would be featured on a label for the novelty chocolate cigars that Milton S. Hershey produced as "A Delicious and Harmless Smoke" at his Church Street confectionary in Lancaster, Pennsylvania before he began his renowned chocolate company in Derry Township (now more commonly known as Hershey, Pennsylvania).

Chocolate had been an essential constitutional drink used in earlier military encounters. Both Union and Confederate troops during the U.S. Civil War used chocolate for fortification as well as for part of their medical regimens according to widely scattered reports in soldiers' diaries, letters and official government documents. Napoleon frequently ate chocolate when he needed a quick energy boost. Both British and Colonial American provincial armies prepared rations of chocolate in the French and Indian (or Seven Years') War in North America.[44] Stretching much further back, even Aztec warriors were issued cacao tablets or wafers to be used for sustenance and energy during their military encounters.

Perhaps the best known story of chocolate, health and war efforts is that of U.S. Captain (later Colonel) Paul Logan who in 1934, on behalf of the Quartermaster Subsistence Laboratory, reintroduced chocolate's use as an emergency soldier ration. In 1937, he appointed Pennsylvania's Hershey Company to purposefully diminish the good taste of their product and create a 4-ounce chocolate bar for troops to use only in emergency. Each bar contained chocolate, sugar, powdered skimmed milk, cocoa butter, vanillin, oats and Vitamin B. Three of these individually foil wrapped 600 calorie "Logan bars" were distributed to GIs as emergency D Rations. As needed in emergency conditions, the caloric content of this D Ration could sustain a solider for a day.

Figure 5.16 American GI with a Chocolate Bar in his Pocket, from *The Story of Chocolate* (1960).

First-hand accounts readily attest that these rations frequently kept soldiers alive when no other food was available. In short, "Many a soldier owed his life to a pocket chocolate bar".[45] In May 1943, Hershey's introduced a more tasty "Tropical Bar" that was designed to withstand the more extreme heat of the Pacific Theatre of Operations (Figure 5.16).

During the post WWII period, similar energising ration kits were included on all Coast Guard vessels as well as on many commercial and pleasure cruise ships.[46] During Operation Deepfreeze (1956–57), chocolate was supplied as an energy restorative to U.S. Navy and Air Force troops in Antarctica. Later, Hershey's developed special "Desert Bars" as rations used during the Persian Gulf War (1990–1991).

Chocolate bars and cocoa drink mix were also heavily consumed for health and sustenance on major explorations. When finding himself "very unwell", William Clark, while on expedition with Meriwether Lewis, "derected [sic] a little chocolate ... prepared of which I drank about a pint and found great relief". At the turn of the 20th century, Cadbury's provided 3500 pounds of chocolate for the Discovery Expedition, Britain's National Antarctic Expedition (1901–04), and J.S. Fry and Sons supplied Robert Falcon Scott with chocolate on what turned out to be his fateful Antarctic Expedition (1910–1913). The chocolate, as revealed by recent photographs of the galley area in Scott's Terra Nova Expedition (1911–1912) Hut in McMurdo Sound, Antarctica, as well as that stored in the Discovery Hut built during the earlier British National Antarctic Expedition, out-survived the explorers and remains visible today.

On 26 July 1971, Hershey rations also went to the Moon aboard Apollo 15. Chocolate's benefits in terms of nutrition and in improving health were provided to astronauts in the form of chocolate pudding and "space brownies". According to Norah Smaridge, the pudding was prepared when water from a squirt gun was injected into a sealed plastic film bag that contained a freeze-dried chocolate bar. Astronauts then "kneaded the mixture, and squeezed out the chocolate pudding into their mouths". "Space Brownies" had been "compressed into small cubes and covered with edible plastic". They were "eaten in one bite" in order "to avoid the danger of crumbs flying around in their space capsule".[47] According to the Hershey jingle,

> Today, chocolate covers the globe.
> It goes with astronauts into space.
> And when champs chomp a bar for fast energy,
> They usually win the race![48]

Whether in a foot race or the space race, chocolate was touted as the essential ingredient for winning. Just as chocolate rocketed to new planetary heights, reinvigorated research efforts into its potential health benefits were taking off here on Earth.

REFERENCES

1. Teresa Dillinger, Patricia Barriga, Sylvia Escárcega, *et al.*, "Food of the Gods: Cure for humanity? A cultural history of the medicinal and ritual use of chocolate", *JN The Journal of Nutrition*, 2000, **130 Supplement**, pp. 2057S–2072S, p. 2065S.
2. Teresa Dillinger, Patricia Barriga, Sylvia Escárcega, *et al.*, "Food of the Gods: Cure for humanity? A cultural history of the medicinal and ritual use of chocolate", *JN The Journal of Nutrition*, 2000, **130 Supplement**, pp. 2057S–2072S, p. 2067S.
3. J. F. Beale Jr., "Cocoa of to-day and yesterday", *Confectioners' Journal*, 1906, **32**, pp. 84–85.
4. Edith C. Williams, *A Bibliography of the Nutritive Value of Chocolate and Cocoa with Quotations and Summaries*, prepared

for the Hershey Chocolate Company by The American Food Journal Institute, Hershey, PA, [1925], p. 40.

5. For brief overviews of the persistent use of chocolate as medicine in this era, see J. G. L. Burnby, "Pharmacy and the cocoa bean", *The Pharmaceutical Historian*, 1984, **14**, pp. 9–12 and Donatella Lippi, "Chocolate and medicine: Dangerous liaisons?", *Nutrition*, 2009, **25**, pp. 1100–1103.
6. Symonds, as cited by Brandon Head, *The Food of the Gods: A Popular Account of Cocoa*, George Routledge & Sons, London, E. P. Dutton, New York, 1903, p. 3.
7. Faussett, as cited by Brandon Head, *The Food of the Gods: A Popular Account of Cocoa*, George Routledge & Sons, London, E. P. Dutton, New York, 1903, p. 22.
8. For comparative historical overviews of these beverages, see Peter B. Brown, *In Praise of Hot Liquors: The Study of Chocolate, Coffee and Tea-Drinking 1600–1850*, York Civic Trust, York, England, 1995 and Jean Maurice Biziere, "Hot beverages and the enterprising spirit in 18th-century Europe", *The Journal of Psychohistory*, 1979, **7**, pp. 135–145.
9. Jean Anthelme Brillat-Savarin, *M. F. K. Fisher's Translation of The Physiology of Taste or Meditations on Transcendental Gastronomy*, North Point Press, San Francisco, 1986, pp. 110–111.
10. A claim further noted in Dr Frederick William Pavy's renowned *Treatise on Food and Dietetics: Physiologically and Therapeutically Considered*, Henry C. Lea, Philadelphia, 1874.
11. Brandon Head, *The Food of the Gods: A Popular Account of Cocoa*, George Routledge & Sons, London, E. P. Dutton, New York, 1903, p. 22.
12. A. L. Benedict, *Practical Dietetics*, G. P. Engelhard and Co., Chicago, 1904 recommended a mixture of chocolate and coffee to provide the beneficial stimulant effect that some patients needed. Such mixtures remain popular in coffee shops today.
13. For these citations and further insight into Fanny Farmer and other American cookery book authors' discussion of the nutritional value of chocolate, see Deanna Pucciarelli, "Chocolate as Medicine: Imparting Dietary Advice and Moral Values Through 19th Century North American Cookbooks", in *Chocolate: History, Culture, and Heritage*, eds., Louis Evan Grivetti and Howard Yana Shapiro, John Wiley & Sons, Hoboken, N. J., 2009, pp. 115–126.

14. Marcia Morton, *Chocolate: An Illustrated History*, Crown, New York, 1986, p. 37. François Massialot's 1734 work, *Nouvelle Instruction pour les Confitures, les Liqueurs, et les Fruits*, Paris, 1734, described the use of solid chocolate in the forms of candy and biscuits as well as in a drinkable form.
15. Marcia Morton, *Chocolate: An Illustrated History*, Crown, New York, 1986, p. 49. Eugene Pelletier and Auguste Pelletier, *Le Thé et le Chocolat dans l'alimentation Publique aux Points de Vue Historique, Botanique, Physiologique, Hygiénique, Économique, Industriel, et Commercial*, Le Compagnie Française des Chocolats et des Thés, Paris, 1861 also wrote about chocolate as the best substance, especially for women and children, whereby to overcome those long intervals between meals.
16. Marcia Morton, *Chocolate: An Illustrated History*, Crown, New York, 1986, p. 49.
17. Brandon Head, *The Food of the Gods: A Popular Account of Cocoa*, George Routledge & Sons, London, E. P. Dutton, New York, 1903, p. 15. 10 George III, c. 10.
18. William Hughes, *The American Physitian or A Treatise of the Roots, Plants, Trees, Shrubs, Fruit, Herbs &c. Growing in the English Plantations in America: Describing the Place, Time, Names, Kindes, Temperature, Vertues and Uses of them, either for Diet, Physick, &c. Whereunto is added A Discourse of the Cacao-nut Tree, and the use of its Fruit; with all the ways of making of Chocolate. The like never extant before*, J. C. for William Crook, London, 1672, section on "Another Way of Making Chocolate" in the chapter "Of the Cacao-Tree and Fruit".
19. "Cocoa and Its Adulterations", *Lancet*, 1851, **1**, pp. 552, 608, 631.
20. Robin Dand, *The International Cocoa Trade*, 2nd edn, CRC Press, Boca Raton, FL; Woodhead Pub., Cambridge, England, 1999, p. 11.
21. J. Othick "The Cocoa and Chocolate Industry in the Nineteenth Century", in *The Making of the Modern British Diet*, ed. Derek Oddy and Derek Miller, Croom Helm, London and Rowman and Littlefield, Totowa, N.J., 1976, p. 87.
22. *Peterson's Magazine* (1891), p. 269, as cited by Laura Pallas Brindle and Bradley Foliart Olsen, "Adulteration: The Dark World of 'Dirty' Chocolate", in *Chocolate: History, Culture, and Heritage*, eds. Louis Evan Grivetti and Howard Yana Shapiro, John Wiley & Sons, Hoboken, N. J., 2009, pp. 625–634, p. 626.

For insight into food fraud history in general, see Bee Wilson, *Swindled: The Dark History of Food Fraud, from Poisoned Candy to Counterfeit Coffee*, Princeton University Press, Princeton, NJ, 2008.

23. W. Clarke Saunders, "Adulteration of cocoa and chocolate", *Confectioners' Journal*, 1895, **21**, pp. 64–65, 64.
24. Philip Porter Gott and L. F. Van Houten, *All About Candy and Chocolate: A Comprehensive Study of the Candy and Chocolate Industries*, National Confectioners' Association of the United States, Chicago, 1958, p. 23. By the 1910s, biochemical analysis and ultraviolet rays were used to identify the adulterated substances within chocolate. R. Wasicky and C. Wimmer, "Eine neue methode des nachweises der schalen im kakao", *Zeitschrift für Untersuchung der Nahrungs-und Genussmittel*, 1915, **30**, pp. 25–27.
25. "*Good Nutrition Makes Good Sense*, Hershey Foods Corporation Collection, Accession 87006, Box B-11, Folder 40, 1982.
26. Russian chemist Alexander Woskresensky isolated and identified theobromine from cacao beans in 1841, as described in Alexander Woskresensky, "Über das theobromin", *Liebig's Annalen der Chemie und Pharmcie*, 1842, **41**, pp. 125–127.
27. J. C. Bainbridge and S. H. Davies, "Essential oil of cocoa", *Journal of the Chemical Society, Transactions*, 1912, **101**, 2209–2221.
28. James D. McMahon, *Built on Chocolate: The Story of the Hershey Chocolate Company*, General Publishing Group, Los Angeles, CA, 1998, p. 72.
29. James D. McMahon, *Built on Chocolate: The Story of the Hershey Chocolate Company*, General Publishing Group, Los Angeles, CA, 1998, p. 72.
30. Bernarr Macfadden, *Hershey, The Chocolate Town*, Hershey Chocolate Company, Hershey, PA, ca. 1923, pp. 26–27.
31. *The Story of Chocolate and Cocoa: With a Brief Description of Hershey, "The Chocolate and Cocoa Town" and Hershey "The Sugar Town"*, Hershey Chocolate Corp., Hershey, PA, 1926, p. 11.
32. Arrowroot was the nutritious starch obtained from the root of the *Maranta arundiacea* plant.
33. G. A. R. Wood and R. A. Lass, *Cocoa*, Longman, London & New York, 1985, p. 5 For a brief historical overview of the

homeopathic use of chocolate, see Deanna L. Pucciarelli and Louis E. Grivetti, "The medicinal use of chocolate in early North America", *Molecular Nutrition & Food Research*, 2008, **52**, pp. 1215–1227, esp. pp. 1222–1223.
34. Marcia L. Pelchat and Gary K. Beauchamp, "Sensory and Taste Preferences of Chocolate", in *Chocolate and Cocoa: Health and Nutrition*, ed. Ian Knight, Blackwell Science, Oxford, England, 1999, p. 316.
35. Jeremy Sherr, *The Homoeopathic Proving of Chocolate*, [B. Jain Publishers, n.p., 1993], p. 6.
36. J. Othick "The Cocoa and Chocolate Industry in the Nineteenth Century", in *The Making of the Modern British Diet*, ed. Derek Oddy and Derek Miller, Croom Helm, London and Rowman and Littlefield, Totowa, N.J., 1976, pp. 77–90, p. 87.
37. Harold McGee, *On Food and Cooking: The Science and Lore of the Kitchen*, Charles Scribner's Sons, New York, 1984, p. 401.
38. Ovaltine was reintroduced to an entire new generation after being featured in that classic holiday movie and now musical, "The Christmas Story". The Clark Bar is still readily available in shops throughout Clark University in Worcester, Massachusetts.
39. "Historicus" [Richard Cadbury], *Cocoa: All About It*, S. Low, Marston & Co., London, 1892.
40. "Chocolate and cocoa", *Scientific American*, 1918, **86**, p. 232.
41. Anthony M. Sammarco, *The Baker Chocolate Company: A Sweet History*, History Press, Charleston, SC, 2009, pp. 65–66.
42. Bonnie Busenberg, *Vanilla, Chocolate & Strawberry: The Story of Your Favorite Flavors*, Lerner Publications, Minneapolis, 1994, p. 68.
43. Armies on the march in the 19th century also greatly benefited from chocolate, so I. Vives reported in "Del arroz, pastas, garbanzos y chocolates, considerados como elementos del regimen alimenticio de los enfermos militares", *Gaceta de Sanidad Militar*, 1882, **8**, p. 176.
44. During the War of 1812, chocolate was provided for members of the Royal Navy and the Provincial Marines in British North America.
45. Chocolate Manufacturers Association of the U.S.A., *The Story of Chocolate*, Chocolate Manufacturers Association of the

U.S.A., 1960, p. 8. See also [Hershey Community Archives], "Hershey's chocolate and the war effort", *Call to Duty*, 2011, **6**, p. 8. Milton S. Hershey turned the patent for these ration bars over to the U.S. Government. Beyond their health benefits, WWII chocolate bars also undertook something akin to the monetary value that the Mesoamericans had once given their precious beans. During the war, chocolate became central to the "barter equation" with the "GI holding out a chocolate bar, [and] the fräulein nodding *ja*". Marcia Morton, *Chocolate: An Illustrated History*, Crown, New York, 1986, p. 125, summarised this in terms of chocolate as "a symbol of the crassness brought out by war's degradation". For more on the administrative history of WWII chocolate consumption, see James J. Cook, *Chewing Gum, Candy Bars, and Beer: The Army PX in World War II*, University of Missouri Press, Columbia, MO, 2009.

46. Donald G. Mitchell, *The Chocolate Industry*, Bellman, Boston, 1951, p. 44, a research chemist for Baker's Chocolate Company noted after WWII that, "[p]rior to the war Germany was internationally known for its design and fabrication of chocolate processing equipment", but that "[i]n the present state of affairs it may be a long time before Germany will again be manufacturing chocolate processing machinery".
47. Norah Smaridge, *The World of Chocolate*, J. Messner, New York, 1969, pp. 83–84.
48. "*Hershey, Color It with Happy*", Hershey Foods Corporation Collection, Accession 87006, Box B-11, Folder 32, ca. 1968–1977, np.

CHAPTER 6

Modern Chocolate Science and Human Health

> Throughout history cocoa and chocolate were beloved for their health-giving properties, but for much of the twentieth century, chocolate was demonized as a high-fat, high-calorie food, offering no more than pleasure. All of that changed about a decade ago when the first news reports came out that cocoa contains plant compounds that may be good for you.
>
> Shara Aaron and Monica Bearden,
> *Chocolate: A Healthy Passion* (2008)

Not all of chocolate's reputed medical benefits promoted in 20th century ads were new. However, it was at this time that some products containing ingredients derived from the cacao bean gained unprecedented popularity. Cocoa butter, for example, known to the Mesoamericans for its wound-healing abilities, increasingly appeared in ads for ointments, suppositories, pessaries and pomades.[1] This was also a key constituent of the chocolate paste that the Indians of Nicaragua smeared over their faces "as a way of enhancing their good looks".[2] Additionally, this product was readily promoted as an emollient for skin massages as well as for healing chapped lips and, in nursing women, as a remedy to heal sore and cracked nipples. Its use in suntan lotion became popular in the late 20th century. For individuals

Chocolate as Medicine: A Quest over the Centuries
Philip K. Wilson and W. Jeffrey Hurst

Published by the Royal Society of Chemistry, www.rsc.org

unable to digest cocoa fat, cocoa-shell infusions were recommended as an analeptic (*i.e.*, restorative) tonic.

Other medical preparations have long relied upon chocolate as a flavoring agent. The harsh taste of many drug ingredients has historically been a primary reason for patients' noncompliance regarding their health care providers' recommendations. Distinct from its other health uses, chocolate has been a favored excipient for centuries to mask the unpleasant, repugnant tastes of such medications. Syrups prepared with chocolate have long been added to enhance the flavor of tonics and elixirs. Throughout the 1800s, cocoa powder was included in the mixture of ingredients compressed into tablets to mask the taste of such active agents as agaracin, aloin, arsenious acid, calomel, chalk, gray powder, phenolphthalein, podophyllum, quinine, santonin and terpin hydrate.

Among chocolate's most widely promoted use as a flavoring agent, at least in recent years, is its inclusion in the senna-containing Ex-Lax – "The Chocolate Laxative". Max Kiss, a Hungarian-born pharmacist working in Brooklyn, New York, initially concocted a chocolate-flavored phenolphthalein laxative that he called Bo-Bos. One day, Kiss was reading a Hungarian-language newspaper report of a deadlock in that country's parliamentary debate. In Hungarian, such deadlocks were colloquially known as an "ex-lax". Ah, "Ex-Lax" would also serve as a fitting name for his "Excellent Laxative", so Kiss surmised, and in 1906, he launched both the product and the company under that name.

In addition to promoting chocolate's use to mask the harsh taste of medicines, some companies also featured it as an ideal agent to "temper" the "irritant actions" of their commercial laxatives and purgatives. Sterling Products, for instance, produced Cascarets that contained the laxative cascara, though they also touted chocolate as an "active ingredient" within their compound.[3] One downside of using chocolate in laxative preparations is that its delectable taste has led to considerable laxative overuse and abuse.

In earlier periods, we find chocolate being widely used as a vehicle for easing the administration and delivery of iron-containing medicines. Iron salts or iron filings had been prepared in the Old World since the first Spanish reports promoting this remedy were received from the New World. These reports also indicate that cacao was commonly mixed with these ingredients, typically by dissolving it in iron water, to produce what were called ferruginous chocolate drinks. August Saint-Arroman, who later promoted chocolate as a form of medical treatment especially for the weak and aged in his *Coffee, Tea & Chocolate: Their Influences upon Health* (1846), noted that ferruginous chocolate held special benefit for "women who are out of order [*i.e.* menstrual irregularities], or [who] have the green sickness [chlorosis, anemia]".[4]

For centuries, chocolate also gained popularity as both a vehicle for administering poisons as well as on occasion a reputed natural antidote to poisons. In Chiapas, in the province of Soconusco, Mexico during the mid-1600s, Don Bernardino de Salazar, Bishop of Chiapas threatened excommunication upon a group of ladies who partook of chocolate during church services in order to help them from growing weak and faint. From the Bishop's view, his Masses and sermons were interrupted as the ladies' maids brought them freshly prepared hot chocolate throughout the service. The women "slighted him with scornful and reproachful words", and they left the Cathedral to worship in the cloister churches.[5] Shortly thereafter, the Bishop died – reputedly after drinking his own daily cup of chocolate that had been poisoned, presumably by the outcast ladies of Chiapas.[6] A scandal reputedly erupted following the Bishop's death, eventually giving rise to the proverb: Beware of the Chocolate of Chiapas. In another instance,

> one morning, the valet-de-chambre of Frederick the Great, King of Prussia, carried in the King's chocolate as usual; but in presenting it, his resolution failed him, and the King remarked [on] his extraordinary confusion. "What is the matter with you? ... I believe you mean to poison me". At those words [the valet's] agitation augments; he throws himself at the feet of the monarch, avows his crime, and begs his pardon. "Quit my presence! Miscreant", answered the King; and this was all his punishment.

However, from that morning on, Frederick "gave a little of it to his dogs" and watched them before he drank of it himself.[7] Pity that it was unknown at the time that dogs lack an enzyme to properly digest chocolate and thus a considerable bolus of chocolate in any form might well prove fatal to them.[8]

One last notable case (though possibly apocryphal) involved another Emperor, Napoleon. In August 1805 while in Lyon, the Corsican Pauline Riotti, a former mistress of the Emperor, reputedly poisoned his cup of chocolate. When a cook noted Riotti acting suspiciously, he alerted Napoleon. The plot was revealed, Riotti drank the cup of chocolate that she had prepared, and she perished. The cook was rewarded with a pension and the Légion d'Honneur.[9]

In the 18th century, chocolate was widely used for "female complaints" similar to the use of Lydia Pinkham's Compound in a later era.[10] In the 19th century, chocolate-flavored medications became increasingly targeted for children's use as well. To fight intestinal worms, Robert Gibson and Sons advertised that the "medical ingredients" of their Penny Chocolate Worm Cakes were "thoroughly amalgamated with pure chocolate so that, in point of fact, they became a delicious Sweetmeat". Indeed, children found these cakes to be an agreeable form of taking "opening medicines" (mixtures typically containing calomel, cinnamon, croton oil and santonin) whereby the intestinal worms were expelled. Chocolate flavored anti-worm products, including Combantrin, were prepared and distributed by drug manufactures through the mid-20th century. Chocolate was also added to bran at this time in order to improve the taste (and boost the sales) of this commonly used fiber product. Post's Bran Chocolate, a "Delicious Health Confection", was marketed to parents as "Bran in Candy Form" for children in the mid-1920s. Post's chief rival, Kellogg's, deemed chocolate to be poisonous to the system.[11] Using chocolate as a flavorant was also behind its inclusion in Castrophene, Bordon's Hemo and Bosco Chocolate Syrup, whose 1951 advertisements shared the perceived benefits from the viewpoints of both a child and a pediatrician.

Though the name *chocolat de salud* (health chocolate) frequently appeared in descriptions of various chocolate-based medicines, during the 19th century, a wide range of drug concoctions used specifically designed chocolate or cocoa nomenclature in order to help promote their sales.[12] One finds, as Graziano has uncovered, advertising for amber chocolate, tonic chocolate, binutritive chocolate of chicken broth, chocolate of pepsonized meat, tar chocolate, healthy chocolate of oats and albumin chocolate.[13] "Headache Chocolate" was sold, as was chocolate digitalis for

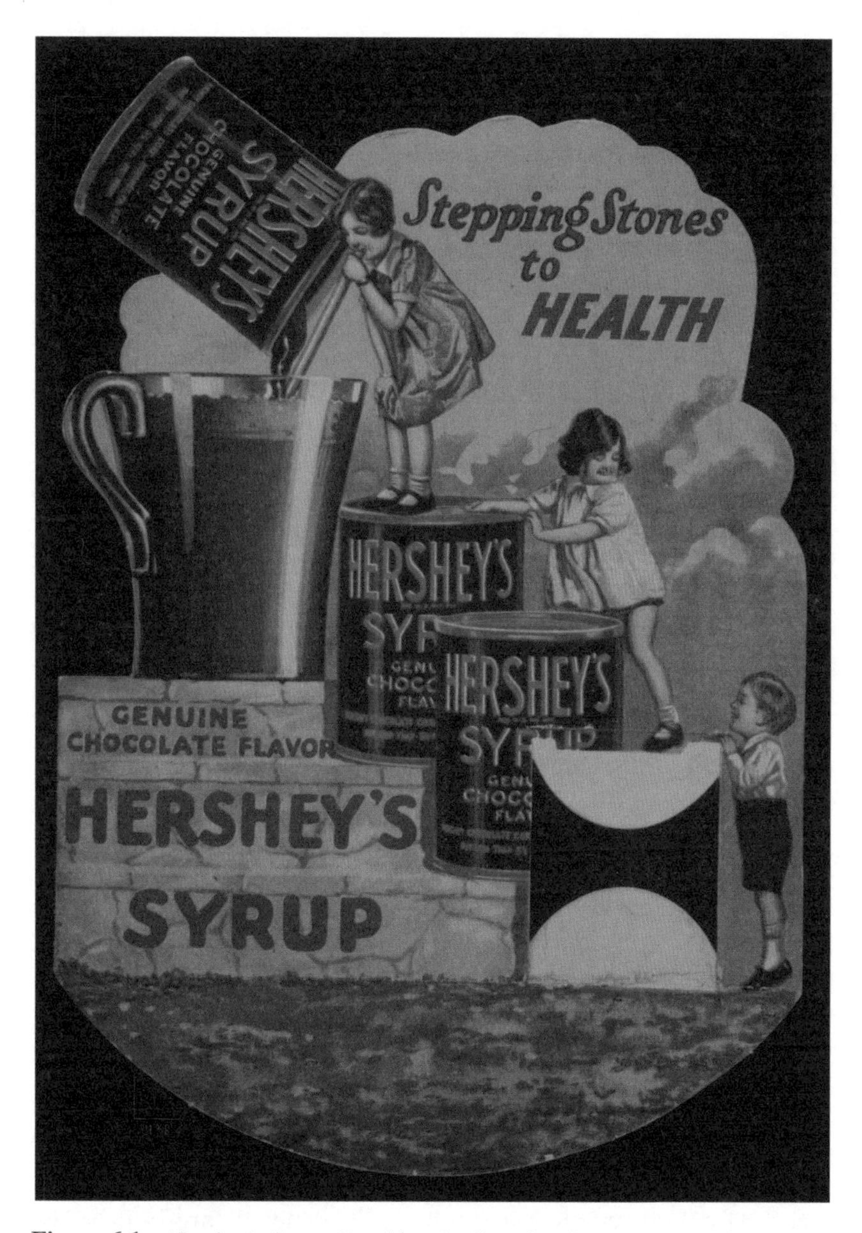

Figure 6.1 Hershey's Syrup Provides the Stepping Stones to Health, Point of Purchase Display (1934–1940).
(Courtesy of Hershey Community Archives, Hershey, Pennsylvania, USA).

heart patients, acorn chocolate to relieve the glands, and products including pectoral chocolate of osmazone, pectoral and eutrophic chocolate, Iceland moss chocolate, and pectoral chocolate of Balsam of Tolu were readily available for patients with chest complaints. Those with stomach ailments could select from such products as stomachic chocolate of Catechu, digestive chocolate of Vichy salts, and stomachic chocolate of cinchona. McVitie's still sells a dark chocolate stomachic product called "Digestives", whereas Cadbury continues to offer their similarly targeted Milk Chocolate Digestive Biscuits. We have previously noted that chocolate was long reputed to promote venery, but it was also added to popular (and top-selling) preparations that were part of venereal disease treatment, including antisyphilitic chocolate with mercuric chloride or with Balsam of Peru, or the Depurative Chocolate with potassium iodide, or as chocolate tablets prepared with antimony or guaiacum. "Diabetic" chocolates were also available at least as early as the 1910s, though on analysis, marketing one German product as such "appear[ed] to be irrational", one contemporary noted, for this chocolate's manufacturer prepared it with a "disproportionately high percentage of starch and cellulose" all the while knowing that introducing these "carbohydrates into food for diabetics should be avoided as much as possible".[14]

During the early 20th century, the Eli Lilly Drug Company produced an entire line of chocolate-flavored medical products. In addition to their most popular Coco-Quinine ("Quinine Sulphate suspended in a chocolate-flavored syrupy medium which masks the [bitter] taste"), they also sold Coco-Santal Oil, Coco-Tablets of Calomel, a "Palatable" Coco-Emulsion of Cod Liver Oil and a Coco-Cordial. Through the 1930s, all major pharmaceutical manufacturers offered chocolate-covered pills, tablets and lozenges as an alternative to their sugar-coated or gelatin-coated varieties. After that time, cacao was still used in the preparation of "medicated chocolates" as a flavorant for sulfa drugs, antibiotics and vitamins. As far back as the 1830s, pharmacies sold carbonated water as a therapeutic agent, often flavoring it with chocolate. Today, many pharmacists still offer the addition of chocolate-flavored syrup to medications they compound, especially those prepared for children. Many dentists also continue to offer their patients chocolate-flavored toothpaste and chocolate-flavored fluoride gel or foam treatments during their office visits.

By the mid-20th century, a new type of evidence was beginning to be commonly used in support of chocolate's reputed medical benefits. These new claims conformed with the concurrent changes seen in the biomedical representations of the human body. Early in the century, writings like that of the popular U.S. medical authority and Kansas City physician and medical school professor Logan Clendening's *The Human Body* (1927) as well as his *The Care and Feeding of Adults, With Doubts About Children* (1927) and his later *The Balanced Diet* (1936) began to focus general reading audiences upon the body's biochemical makeup. Similar depictions of chocolate's biochemical composition also began to appear at this time. Such an account appeared in *The Story of Chocolate & Cocoa* (1926) which claimed that,

> Milk is Nature's best food. It contains a considerable percentage of protein, which is a body-building material. With its fats, it provides heat and energy. Again, cream of milk fat is a nerve food, containing that most important of all vitamines [sic], fat soluble "A". Milk is at the same time our best lime food and in general our best mineral food. It is chemically a perfectly balanced food, upon which one never will or never can develop an acid-saturated state of tissues of the body, because milk contains the base-forming elements which neutralize acids formed in the system. Not only is milk the best lime food, but it is the best of all foods for supplying the organic salts of potassium, phosphorus, magnesium, sodium, chlorine, iodine, silica and those other materials found in the human body in small quantities, but most vitally needed in small quantities.[15]

Twenty-five years later, Samuel Hinkle, then Chief Chemist of the Hershey Company, was still touting chocolate's nutritional benefits. All "activities of the human body" were known to "require a constant expenditure of energy" and an "interchange of material". Chocolate products – particularly Hershey's chocolate products – were offered to the public "with the knowledge that they contain the highest grade ingredients prepared under rigid sanitary conditions and ... [were] made of the finest [chocolate] that can be made". The value-added benefit being that they were also "sources of highly concentrated food energy" such that chocolate products have earned a "rightful place" alongside "all well-known and well-prepared foods" (Figure 6.1).[16]

6.1 EXPERIMENTALLY SUPPORTED BENEFITS FROM CHOCOLATE CONSUMPTION

Though Sir Francis Bacon's recommendations from 1620 had long influenced scientific thought, a reinvigorated experimental method emerged in medical practices during the second half of the 20th century. Within this new movement, quests for experimentally derived evidence were undertaken to support claims of chocolate's medical benefits. Biomedical science began to experimentally assess chocolate's potential for alleviating medical disorders just as it did all other pharmaceutical products. Randomised control trials of increasingly complex design based at multiple clinical sites were used to identify standards of normalcy and degrees of difference.[17] By the close of the century, claims for chocolate's medical benefits were supported by a growing "science" of chocolate. As was noted, "Modern science, would, in time, add proof to what the Aztecs knew and practiced. Chocolate does confer energy".[18] Though anecdotal evidence persisted in an attempt to convince consumers, biomedical evidence was becoming much more prevalent throughout all types of advertisements.

In the late 1900s, claims of chocolate's benefits (derived primarily from the nib within the cacao bean) focused upon its richness in carbohydrates and fat. Chocolate's natural phytonutrient flavonoid phenolics were known to prevent the rancification of fat, thereby diminishing the need to add preservatives to chocolate each of which might contribute its own health risks.[19] The plant-derived, saturated stearic acid fats present in cacao were fortunately not those guilty of increasing cholesterol levels. Considerable recent investigation has centered around a particular flavonoid, epicatechin.[20] Following chocolate consumption, epicatechin has been found to promote cellular antioxidation that decreases LDL (low-density lipoprotein, aka "bad") cholesterol activity, thereby delaying the onset or progression of atherosclerosis and arteriosclerosis. As American Cocoa Research Institute-funded work from the lab of Penn State University

Distinguished Professor of Nutrition Penny Kris-Etherton uncovered over a decade ago, dark chocolate also increases HDL (high-density lipoprotein, aka "good") cholesterol levels.[21] Other researchers have since shown that chocolate initiates antiplatelet activity, thereby reducing plaque formation and platelet-clotting properties that impede blood flow.[22] Flavonoids are known to stimulate blood flow in brain, hands and legs due to regulation of nitric oxide synthesis. Dark (high cacao concentrated) chocolate also works to reduce blood pressure by promoting blood-vessel dilation. Not surprisingly, such medical literature reports quickly appeared in all news media formats. As reported in a 1996 *Chocolatier* magazine editorial, "All of a sudden 70 percent cacao solid chocolate bars ... [became] the rage. Two years ago you couldn't give them away".[23]

Similar to assessments undertaken on any drug, possible adverse reactions to chocolate have been investigated. Though some preliminary reactions have been repeatedly suggested, to date little conclusive biomedical evidence is available. In 1982, a Hershey's promotional brochure distanced its company product from acne and cavities, noting the work of specific authorities as their evidence. Dr Joseph Fries, a "nationally recognized allergist" was cited as having "reviewed over 150 research studies on the effects of chocolate on humans". In opposition to much public belief, Fries concluded that "chocolate is irrelevant to the causation of acne". Further, noting *in vitro* testing at the Massachusetts Institute of Technology and the Forsythe Dental Center in Boston, chocolate was found to contain a "heat-resistant, water-soluble component [tannin] that blocks production of an enzyme which causes dental plaque, one of the first steps in tooth decay". Cocoa butter within chocolate products was found to coat the teeth, thereby countering the cariogenic (dental caries or decay promoting) effect of chocolate's high sugar content by actually preventing tooth decay. In addition, Hershey's claimed that "scientists recognize that cocoa butter in [chocolate] milk helps the food clear the mouth quickly".[24] Still, researchers continue seek to more clearly elucidate chocolate's role in affecting headaches, bone density, kidney stones, allergic reactions, insulin sensitivity, acne and esophageal reflux. Concern over the possibility of transmitting the active ingredients of chocolate via breast milk is also being investigated.

6.2 CHOCOLATE ON THE BRAIN

A 1906 *Confectioners' Journal* commentary upon a recent *New York Herald* article about "Hysteria and Chocolate" recounted that

> a young woman employed in a chocolate-making establishment became hysterical and was taken to Bellevue Hospital. It is intimated that this form of illness is frequent among chocolate workers. Just what malign effect chocolate may have upon feminine nerves is not made clear. [Given this to be a] lone instance, [the report continued that it was more regularly] shown that a box of chocolates has often proven a prompt and efficient remedy for nervous collapse among women.[25]

For decades, high-quality dark chocolate's reputed psychoactive and possible aphrodisiac attributes have been linked to its high concentration of the methylxanthin alkaloid stimulant, theobromine. Late 20th-century investigators found that when people become infatuated or fall in love, the brain levels of phenylethylamine (PEA) – the reputed "love drug" – are increased. In turn, PEA enhances the release of endorphins, what Aaron and Bearden refer to as those "feel-good chemicals".[26] In the early 1980s, chocolate was also found to promote this release, though in relatively small quantities. In Hershey Company research, W. Jeffrey Hurst, *et al.*, compared PEA-containing foods, finding that chocolate promoted only low level release, thereby strongly suggesting that these compounds were most likely not responsible for self-reported desire for chocolate.[27] Chocolate appears to promote the neurotransmitter serotonin release as well, thereby producing calming, pleasurable feelings. Finally, anandamide – a neurotransmitter discovered in 1992 and named after *ananda*, the Sanskrit for "happiness" or "blissfulness" – was found to be structurally similar to marijuana's tetrahydrocannabinol (THC), a compound that, like anandamide, activates cannabinoid neuroreceptors known to change the texture of consciousness. Anandamide is also released following chocolate consumption, likely contributing to the euphoria which many claim that chocolate induces. All of these psychopharmacological alterations may contribute to chocolate's perceived aphrodisiac effects, though the precise interconnections have yet to gain substantial scientific support.

Following WWII, another of chocolate's beneficial claims began to take center stage. Namely, chocolate was being advertised as a means of allaying mental stress. We find this in an enduring advertisement of Fry's Chocolate in which a young lad's face was shown sequentially moving through five sensations – desperation, pacification, expectation, acclamation and finally realisation – depicting anticipated craving-like experiences while longing for a Fry's chocolate bar (Figure 6.2). Concurrently, ads and articles increasingly suggested "chocolatomania" cravings, particularly in women. Mental stress and chocolate craving were highlighted in an advertisement labeled "Martine and Her Problems – Chocolate: Whim or Necessity?" In this ad from a 1955 *Paris-Match Magazine*, we learn the following:

> Often, Martine, you feel like eating a bar of chocolate .... but you don't, because you think you shouldn't – and so you are being unfair to yourself. You don't want chocolate; you need it! You've done the shopping and the housework, and still somehow found the energy to be the perfect hostess. You are exhausted, your body urgently requires energy in

Figure 6.2 Anticipating Fry's Chocolate in 1920s Ad.

> compensation, and you naturally feel the urge to eat chocolate, because it is a balanced food which instantaneously restores the essential elements that you have used up. You deserve it, so why feel guilty about tucking into a bar of chocolate?[28]

Forty years later, women were still reporting similar messages – and more – to their health-care providers.

> I hate cooking, I live on my own, I love chocolate, chocolate makes it quick to prepare a meal, I enjoy this meal thoroughly, eating chocolate requires little work in the kitchen, no cleaning, chocolate eating is satisfying, affordable and gives pleasure, so why should I not live on chocolate?[29]

Over the remainder of the century, inquiries flooded into psychologists regarding what might best be termed, "chocolate psychology". As pop culture scholar and Worcester State College media professor emerita Linda Fuller reminds us, chocolate has gained at least a folk reputation across a wide range of psychological conditions. It has been viewed as

> a cure for agoraphobia, a security blanket, an appeal to the senses, a strategy of control, a means to excite passions and curiosities, an excuse gratification, an antidote to guilt, a step toward intimacy, a cause of dependence, a stress-reliever, a restorer of strength, a reward, or maybe just a simple consolation in a world of contracting opportunities.[30]

Questions also continue to arise over chocolate's reputed addictive nature, some investigators classifying it in terms of emotional rather than physical addiction. Unlike the discussion surrounding other reputed addictions, chocoholism has not been stigmatised as something to be avoided. Such claims of chocoholism, however, are hardy novel. The Jesuit Priest, Josè de Acosta, writing in 1590, noted chocolate to be a "crazy thing" valued in New Spain (*i.e.*, Mexico) by men and "even more" by women who appear to have become "addicted" to the chocolate drink.[31] Integrative Medicine guru, Dr Andrew Weil, and Winifred Rosen addressed chocolate "addiction" in their highly acclaimed, *From*

Chocolate to Morphine: Everything You Need to Know About Mind-Altering Drugs (1983). They claim that "cases of chocolate dependence are easy to find". Most chocoholics are women, they argue, and "many of them crave chocolate most intensely just before their menstrual periods. Women who develop an addictive relationship with chocolate usually eat it in cyclic binges rather than continually, and often say that it acts on them like an instant anti-depressant".[32]

Despite anecdotal evidence that may say "Yes, Yes, Yes", little modern biomedical evidence supports the claim that one can become addicted to chocolate. Since depriving one of chocolate fails to produce scientifically significant signs of withdrawal, it is thereby not technically classed as a physically addictive agent. Furthermore, scientists have not shown a state of dependence regarding chocolate's use.[33] Recent trends tend to avoid language of "addiction" – including "rational chocolate addiction"[34] – in favor of investigations into the "psychological drivers of chocolate consumption"[35], or eating chocolate as an encouraging "habit" that may be "good" and that may have excellent outcomes, including the ability to improve one's positive mental attitude.[36] Such habits may well be driven by needs to re-establish some inner balance or homeostasis. These needs, which certainly seem to have a psychological basis, may also be biologically, evolutionarily and culturally driven as well.[37] Here, the historical context is also of importance, for it was a matter of achieving balance – balance in terms of aligning the four bodily humours – that governed the drive and manner by which the peoples of Mesoamerican cultures first employed chocolate as medicine.

Some have argued that chocolate may pharmacologically stimulate behaviors of compulsive eating, others claim that this finding may just as well be the result of a more generalised aesthetic craving for the sweetness and oily richness and complete orosensory experience that chocolate provides. Indeed, chocolate's rich natural complexity – a complexity that rivals any other food – makes identifying the actual source of perceived cravings or chocoholism exceedingly difficult.[38] In other words, complexities in distinguishing causal factors regarding qualities that include preferring, liking and craving chocolate may well be inexorably tied to the complexities of the substance itself. Despite the lingering concern over categorising chocolate cravings in the bioscience literature, many will likely follow the folk wisdom that Khodorowsky and Robert shared in *The Little Book of Chocolate*: whether addictive, habitual or merely craved, chocolate "will always be assured a place of honor in the pharmacopoeia of pleasure".[39]

6.3 CHOCOLATE – A PANACEA?

One offshoot of 20th-century medical reform was the extensive research expended toward drug design and development. By the end of the century, drugs dominated the medical marketplace. Pharmaceutical industries have grown into mega-businesses that commandeer physicians' polypharmacy practices. Still, for many health care providers a paradigm of polypharmacy is not a panacea.[40] The growing disenchantment with polypharmacy is strikingly clear in consumer health movements for more holistic, integrative health care. Patients' skepticism over the benefits derived from traditionally prescribed remedies alone is evidenced by the increasing demand for more natural medicine.

What could be more natural than chocolate? As the nomenclature *Theobroma cacao* suggests, we have long viewed chocolate as a food of the Gods. Steadily, authorities have been reinforcing chocolate's potential medicinal benefits. According to Harvard Medical School's Norman K. Hollenberg, the "pharmaceutical industry has spent tens, probably hundreds of millions of dollars in search of a chemical that would reverse ... [or ward off vascular diseases]. And God gave us flavanol-rich cocoa which does that".[41]

With increasing awareness focused upon the cacao content of chocolate products, U.S. chocolate manufacturers during the 1990s, following a trend begun by French chocolatiers Valrhona and Bonnat a decade before, turned their attention to the popular *couverture* preparations of European businesses who staked their reputations upon the production of high-quality chocolate with high cacao and low sugar content. The healthiness of the product now seems to reflect the preservation of as much of the natural cacao content and flavor as possible.[42] A concurrent interest has arisen in using only pure and certified organic cacao, an interest that harkens back to the rage over pure chocolate products of the mid-19th century. Similar historical throwbacks are also noticed in the current branding of "diet" chocolates much like that of the popular "dietetic" chocolates of yesteryear. However, rather than advertising chocolate as providing complete nourishment as in the past, modern-day ads focus more upon its value as supplemental nutrition.

As chocolate's perceived benefits have gained substantial support from experimental biomedicine over the past two decades, the summary presented below provides a snapshot of current thought and concerns in this area of research. Although brief in its overview, perhaps in the future this information will provide a helpful backdrop for appreciating which areas have gained even greater support in comparison to this point in time.

6.4 EARLY 21st-CENTURY MEDICAL USE OF CHOCOLATE

More than any other area, researchers have focused upon chocolate's potential impact upon cardiovascular health. Epidemiological results have provided evidence of chocolate's emergence as a "possible modulator of cardiovascular risk".[43] Studies of the Kuna Indians in Panama, a group that consumes large amounts of cocoa daily, have demonstrated that chocolate in the diet has created some measure of protection against hypertension and cardiovascular disease.[44] One 2011 cross-sectional study of nearly 5000 patients in the United States shows an inverse correlation between chocolate consumption and the prevalence of cardiovascular disease.[45] Other cardioprotective results of chocolate were found in independent cohort epidemiological studies performed in the United States, Germany and Sweden.[46]

Results from *in vitro* studies have provided striking and statistically significant evidence that the cocoa-derived[47] polyphenol known as flavanol (also known as flavan-3-ol or catechin) can metabolically diminish the development of cardiovascular disease. Here, most attention has been focused upon atherogenesis, *i.e.*, the process of atherosclerotic plaque development that is an underlying factor shared by many cardiovascular diseases and cerebrovascular (stroke) events.

Chocolate, especially flavanol, has been shown to modify the inflammatory cascade processes linked with the formation and subsequent destabilisation of the plaques that accompany many cardiovascular diseases. Initial work from the mid-1990s looked predominantly at the antioxidant activity of flavanols. Here, these cocoa-derived products were found to assist in the scavenging of free radicals and the chelation of metals in ways that enhanced tissue, cellular and plasma oxidative defense mechanisms, thereby downregulating the overall inflammatory response.[48]

Among the specific areas of chocolate's inflammation modulation abilities being investigated are flavanols' roles in inhibiting the cellular activation and expression of matrix metaloproteins that, themselves, aid in plaque destruction;[49] interfering with the crucial inflammation regulatory response transcription factor known as NF-κB (nuclear factor-kappaB) in leucocytes[50]; combination with cytokines, those key biomolecular mediators that initiate and promote inflammation[51]; and the production of eicosanoids that modulate vascular permeability and the recruitment of immune cells into the vascular wall.[52]

In addition, mononuclear blood cells have been shown to have established greater resistance to oxidative damage 2 hours after healthy individuals consume 45 g of dark chocolate compared to those from individuals who consume flavonoid-deficient white chocolate. This effect is transient, with no significant difference in resistance observable 22 hours after chocolate consumption; however, resistance is re-established following the consumption.[53] Importantly, this finding suggests that consuming small amounts of cocoa at regular intervals throughout the day may well have greater health benefit than consuming a larger amount of cocoa in a single serving.

Another prominent area of research involves examining vascular endothelial cell tone and stability. Among other properties, endothelial cells contribute to the repair of injured areas of the vasculature. Studies have shown that cocoa-derived flavanols have improved the endothelial-dependent dilatory response of the vascular wall as well as enhanced the mobilisation of bone marrow-derived endothelial progenitor cells to areas of vascular injury.[54]

Blood-platelet activation and aggregation are important steps in plaque formation as well as in the thrombogeneic response following plaque destabilisation. Cocoa-derived flavanols have repeatedly been shown to influence these platelet actions by limiting their effects. For example, 6 hours after consuming chocolate, platelet activation was found to have diminished.[55] Longer-term "cocoa supplementation" was found to reduce platelet aggregation to a degree similar to that of taking aspirin.[56]

Consuming a flavonoid-rich diet has been shown to strongly correlate with improved cardiovascular health.[57] Moreover, the effects and mechanisms of action of large doses of cocoa-derived flavonoids on cardiovascular function have received repeated favorable review.[58] Though the results are striking, the amount of flavonoids ingested in most of these studies is far beyond what could be attained through eating or drinking commercially available products and certainly well beyond that consumed historically by Mesoamericans and Europeans, with the possible exception of Moctezuma II who purportedly drank 50 cups of cacao-based beverages per day. Therefore, the few studies that have shown changes in cardiovascular physiology with small and/or chronic cocoa consumption are perhaps of greater interest in relation to effective dosages of both the historical and modern uses of medicinal cocoa.

Other promising cardioprotective mechanisms underlying chocolate's abilities are beginning to be more thoroughly examined through experiments measuring blood pressure, blood oxygenation and blood flow. The most elegant study to date of cocoa's effect on cardiovascular risk has shown that both systolic and diastolic blood pressure are decreased following the consumption of only 6.3 g of dark chocolate per day for 18 weeks in older subjects with mild hypertension.[59] Magnetic resonance imaging (MRI) has shown that consuming a cocoa beverage containing approximately the same amount of flavonoids as 20 g of dark chocolate for 5 days results in improved oxygenation of blood, likely due to increased blood flow, in the brains of healthy young adults.[60] Improvements in coronary blood flow have also been observed 2 hours following the consumption of 40 g of dark chocolate in heart-transplant recipients[61] and in healthy subjects who consumed 45 g of dark chocolate per day for 2 weeks.[62]

When systematic review and meta-analysis of randomised controlled studies as well as of observational studies have been performed, promising results have been reported. For instance, one 2011 report from "Old and New World" chocolate researchers in Cambridge, England and Bogata, Colombia analysed the data gathered from 114 009 humans reported in 4576 references which assess the risk of "developing cardiometabolic disorders by comparing the highest and lowest levels of chocolate consumption".

Their findings strongly suggest a "beneficial association" between higher levels of consumption and reduced risk of developing these disorders. Individuals who were among the highest levels of chocolate consumers were associated with a "37% reduction in cardiovascular disease" and a "29% reduction in stroke" as compared to individuals among the lowest levels of chocolate consumers.[63]

Another systematic review and meta-analysis of randomised, controlled trials including the evaluation of 1106 individuals participating in 24 different studies provided evidence that flavonoid-rich cocoa significantly reduces blood pressure, improves insulin resistance and decreases both total cholesterol and LDL ("bad" cholesterol) in lipid profiles. This study concluded with the encouragement of further studies to "determine whether these findings translate to an improvement in adverse cardiovascular outcomes".[64]

Other disorders that have received considerable attention amongst researchers regarding chocolate treatment include gastrointestinal and respiratory disorders as well as cancer. Cacao extracts have been shown to inhibit the intestinal chloride channels that *Escherichia coli* and *Vibrio cholera* activate, thereby inducing diarrhea.[65] Such finds corroborate Mesoamerican claims for cacao's effectiveness in treating diarrhea. It has also recently been shown that providing 1000 mg of the compound theobromine, equivalent to what would be found naturally in approximately 50 g of dark chocolate, suppressed irritant-induced coughing in healthy subjects.[66] Regarding the cancer-generating (carcinogenic) process, exploration continues regarding the ways in which cocoa-derived polyphenols possibly 1) suppress the overexpression of pro-oxidant enzymes, 2) inhibit the targeted oncogenes involved in cancer-cell proliferation, and 3) induce apoptosis (the process of programmed cellular death).[67]

Other areas of chocolate's potential in contributing to overall bodily and mental health that are being explored include assessing the role of flavonoids in postmenopausal women, in immunodeficient subjects and in patients who are undergoing antiviral treatment for chronic infections.[68] Others are investigating chocolate's potential role in improving cognition, alertness and body coordination due to its ability to modulate cerebral blood flow. Researchers have also recently demonstrated that consuming high-quality, dark chocolate on a more regular basis actually increases lean body mass, lowers body fat and decreases body weight. The mechanisms thought to be responsible for this reaction involve the phytonutrient antioxidants in chocolate acting to stimulate muscle activity thereby increasing metabolism. These researchers' findings that "more frequent chocolate intake is linked to lower BMI [Body Mass Index]" is, at the very least, "intriguing".[69]

Recent systematic literature reviews have assimilated knowledge ranging from decades-old individual claims to multinational studies, subjecting this data to a variety of meta-analysis computation. Among the leading claims is that despite considerable *in vitro* suggestive findings, the "relative paucity of data" from *in vivo* work warrants further attention placed upon performing "rigorous controlled human [*in vivo*] studies with adequate follow-up and with the use of critical dietary questionnaires".[70]

Along these lines, one 2008 literature review[71] proposed the following checklist as a helpful guide to planning future cocoa and chocolate research trials:

1. Where possible, conduct randomised, controlled, crossover, multidose trials.
2. Use well-defined cocoa or chocolate if possible and have industry prepare similar cocoa/chocolate for independent researchers for future studies.
3. Ensure the bioavailability of the active component from its matrix.
4. Use an appropriate control of nonpolyphenol chocolate.
5. Recruit volunteers with at least one nonoptimal biomarker or disease risk factor.
6. Use a dose of cocoa or chocolate that can readily be incorporated into the daily diet, giving appropriate dietary advice to volunteers in terms of balancing overall energy needs.
7. Measure the cocoa/chocolate composition including the polyphenol profile before and after the trial, noting the stability of any storage or batch variations.
8. Ensure that the final publication contains the analytical results along with the appropriate description of analytical methodology.
9. Carefully assess the biological relevance of the chosen biomarker, with special attention to antioxidant biomarkers.

10. Strive for transparency by registering human trials before they start with a recognizable database, *e.g.*, www.clinicaltrials.gov.
11. Attempt to publish null or negative results to enable balancing of the literature and preventing needless duplication of the work. Challenge journals if papers are rejected solely on this basis.

Manuel Rusconi and Ario Conti of the Alpine Foundation for Life Sciences in Olivone, Switzerland have also offered helpful reminders to be taken into consideration when designing future chocolate biomedical studies. Their aims are to help steer around claims that the "wide variation in cocoa processing and in the content and profile of polyphenols make it difficult to determine to what extent the findings about positive effects expressed in different studies, translate into tangible clinical benefits".[72] To further help discern this translation, one of these authors, Ario Conti, joined Rodolfo Paoletti, Andrea Poli and Francesco Visiolo in editing *Chocolate and Health* (2012), a comprehensive book-length account on the health effects of cocoa and chocolate.[73] This work complements that of Ronald Ross Watson, Victor R. Preedy and Sherma Zibadi's edited 40-chapter *Chocolate in Health and Nutrition* (2013). In future well-controlled clinical studies, it would also be of great utility to use amounts of commercially available cocoa that can be easily adapted into diets in order to more precisely understand cocoa's potential benefits in terms of both nutrition and health.

REFERENCES

1. Louis Evan Grivetti, "Medicinal Chocolate in New Spain, Western Europe, and North America", in *Chocolate: History, Culture, and Heritage*, eds. Louis Evan Grivetti and Howard Yana Shapiro, John Wiley & Sons, Hoboken, N. J., 2009, pp. 67–88, p. 81, notes the medical powers of other parts of the Chocolate Tree, including using the bark to treat bloody stools and to reduce abdominal pain; the flowers for cuts and, when mixed with water, to reduce timidity; the pulp to facilitate childbirth; and the leaves as antiseptics and astringents.
2. J. Eric S. Thompson, "Notes on the Use of Cacao in Middle America", *Notes on Middle American Archaeology and*

Ethnology, Carnegie Institution of Washington, Department of Archaeology, 1956, **128**, pp. 95–116, p. 106.
3. Martha Makra Graziano, "Food of the Gods as mortals' medicine: The uses of chocolate and cacao products", *Pharmacy in History*, 1998, **40**, pp. 132–146, pp. 139–140.
4. Martha Makra Graziano, "Food of the Gods as mortals' medicine: The uses of chocolate and cacao products", *Pharmacy in History*, 1998, **40**, pp. 132–146, p. 139. See also Martha Few, "Chocolate, sex and disorderly women in late-seventeenth- and early-eighteenth-century Guatemala", *Ethnohistory*, 2005, **52**, pp. 674–687 on chocolate and "disordered women" in history.
5. Thomas Gage, *The English-American: A New Survey of the West Indies, 1648*, ed. A. P. Newton, George Routledge & Sons, London, 1928, p. 162.
6. Thomas Gage, *The English-American: A New Survey of the West Indies, 1648*, ed. A. P. Newton, George Routledge & Sons, London, 1928, pp. 162–163.
7. *The Mail, or Claypoole's Daily Advertiser*, 26 January 1792, p. 2, as cited in Louis Evan Grivetti, "Dark Chocolate: Chocolate and Crime in North America and Elsewhere", in *Chocolate: History, Culture, and Heritage*, eds. Louis Evan Grivetti and Howard Yana Shapiro, John Wiley & Sons, Hoboken, N. J., 2009, pp. 255–262, p. 258.
8. For brief discussions, see Frederic P. Miller, Agnes F. Vandome and John McBrewster, eds., *Health Effects of Chocolate: Chocolate, Epicureanism, Cocoa, Dark Chocolate, Circulatory System, Anticancer*, Alphascript Publishing, Beau Bassin, Mauritius, 2010, p. 3, and Phillip Minton, *Chocolate: Healthfood of the Gods: Unwrap the Secrets of Chocolate for Health, Beauty and Longevity*, 2011, available through Amazon.com and bn.com, pp. 185–187.
9. *Massachusetts Spy, or Worcester Gazette*, 13 May 1807, p.1, as cited by Louis Evan Grivetti, "Dark Chocolate: Chocolate and Crime in North America and Elsewhere", in *Chocolate: History, Culture, and Heritage*, eds. Louis Evan Grivetti and Howard Yana Shapiro, John Wiley & Sons, Hoboken, N. J., 2009, pp. 255–262, pp. 258–259.
10. Sarah Stage, *Female Complaints: Lydia Pinkham and the Business of Women's Medicine*, W.W. Norton, New York, 1979.

11. Joël Glenn Brenner, *The Emperors of Chocolate: Inside the Secret World of Hershey and Mars*, Random House, New York, 1999, p. 249, In 1837, the United States' Graham Cracker developer and dietary reformer, Reverend Sylvester Graham prohibited the consumption of chocolate, which he considered to be a "vile indigestible substance".
12. In contrast to "Health Chocolate", *cacao añejo* (old cacao) was that deemed to be of inferior quality that would likely prove harmful to consumers and patients.
13. Martha Makra Graziano, "Food of the Gods as mortals' medicine: The uses of chocolate and cacao products", *Pharmacy in History*, 1998, **40**, pp. 132–146, p. 139.
14. Paul Zipperer, *The Manufacture of Chocolate*, 3rd English Edition, Spon and Chamberlain, New York, 1915, p. 310.
15. *The Story of Chocolate and Cocoa: With a Brief Description of Hershey, "The Chocolate and Cocoa Town" and Hershey "The Sugar Town"*, Hershey Chocolate Corp., Hershey, PA, 1926, p. 22.
16. S. F. Hinkle, *Fuel Values of Foods*, Hershey Foods Corporation Collection, Accession 87006, Box B-11, Folder 36, ca. 1936–1949. Hinkle, a Penn State University alum, became President of the Hershey Chocolate Corporation in 1956. In 1963, as a result of a "$50 Million phone call" to then Penn State University President Eric Walker and speaking on behalf of the Hershey Trust Company Board of Directors, Hinkle created the foundation of what is now the Penn State Hershey College of Medicine and M.S. Hershey Medical Center. Founding Penn State Hershey Professor C. Max Lang recently published an historical overview of this medical school's founding. (C. Max Lang, *The Impossible Dream: The Founding of the Milton S. Hershey Medical Center of the Pennsylvania State University*, AuthorHouse, Bloomington, IN, 2010).
17. See, for example, Alvan R. Feinstein, *Clinical Judgment*, Williams and Wilkins, Baltimore, 1967; Kathryn Montgomery, *How Doctors Think: Clinical Judgment and the Practice of Medicine*, Oxford University Press, Oxford, 2005; and Jerome Groopman, *How Doctors Think*, Houghton Mifflin Company, New York, 2007.

18. Adrianne Marcus, "A Brief History of Chocolate", in *The Chocolate Bible*, Adrianne Marcus, G. P. Putnam's Sons, New York, 1979, pp. 27–35, p. 29.
19. Phytochemicals, as Deanna Pucciarelli and James Barrett remind us in "Twenty-First Century Attitudes and Behaviors Regarding the Medicinal Use of Chocolate", in *Chocolate: History, Culture, and Heritage*, eds. Louis Evan Grivetti and Howard Yana Shapiro, John Wiley & Sons, Hoboken, N. J., 2009, pp. 653–666, p. 661, are non-nutritive plant components that are believed to confer a health benefit. Flavonoids are, together with carotenoids and tetrapyrrole derivatives, three of the most important natural pigments. Flavonoids are plant polyphenols frequently found in fruits, vegetables and grains. They are responsible for the color of flowers, fruits and sometimes leaves. The cacao bean is a rich source of polyphenols, the most predominant bioactive forms being the stereoisomers (−)-epicatechin and (+)-catechin that combine into the polymer group of procyanidins, themselves among the least understood of all polyphenols. For a review of the relative bioactivity of these isomers and the effect that cacao bean process has upon their abilities, see Augusta Caligiani, Martina Cirlini and Geraldo Palla, "Cocoa (*Theobroma Cacao L.*) Catechins: Occurrence, Health Effects and Modifications during Processing", in *Chocolate, Fast Foods and Sweeteners: Consumption and Health*, ed. Marlene R. Bishop, Nova Science Publishers, New York, 2010, pp. 231–244, and Monique Lacroix, "Polyphenols in Cocoa: Influence of Processes on Their Composition and Biological Activities", in *Chocolate, Fast Foods and Sweeteners: Consumption and Health*, ed. Marlene R. Bishop, Nova Science Publishers, New York, 2010, pp. 183–197.
20. Monique Lacroix, "Polyphenols in Cocoa: Influence of Processes on Their Composition and Biological Activities", in *Chocolate, Fast Foods and Sweeteners: Consumption and Health*, ed. Marlene R. Bishop, Nova Science Publishers, New York, 2010, pp. 183–197, pp. 185–186, noted that processing chocolate with alkali via the "Dutching" method destroyed the flavonoid content. A century earlier, F. Bordas, *De l'addition de Carbonate de Potassium aux Cacaos*, Paris, 1910 had promoted the addition of alkali as a means of aiding suspension

formation by saponifying and emulsifying the fat, all the while softening the cacao fiber content.

21. P. M. Kris-Etherton, J. A. Derr, V. A. Mustad, *et al.*, "Effects of a milk chocolate bar per day substituted for a high-carbohydrate snack in young men on an NCEP/AHA Step 1 diet", *American Journal of Clinical Nutrition*, 1994, **60**, pp. 1037S–1042S; Ying Wan, Joe A. Vinson, Terry D. Etherton, *et al.*, "Effects of cocoa powder and dark chocolate on LDL oxidative susceptibility and prostaglandin concentrations in humans", *American Journal of Clinical Nutrition*, 2001, **74**, pp. 596–602; and Janet Raloff, "Chocolate hearts: Yummy and good medicine?", *Science News*, 2000, **157**, pp. 188–189.
22. Some cardiologists have also ventured into the cookbook business. For instance, Connecticut practitioner, Robert G. Schneider, together with his wife Joyce, promoted unsweetened cocoa as a "blockbuster drug" in their book, *The Cardiologist's Wife's Chocolate Too! Diet: No Sugar, No Fat, and Luscious*, BooksSurge Publishing, Charleston, SC, 2007.
23. Ruth Lopez, *Chocolate: The Nature of Indulgence*, H. N. Abrams in association with the Field Museum, New York, 2002, p. 119.
24. "*Good Nutrition Makes Good Sense*", Hershey Foods Corporation Collection, Accession 87006, Box B-11, Folder 40, 1982.
25. "Hysteria and chocolate", *Confectioners' Journal*, 1906, **32**, p. 69.
26. Shara Aaron and Monica Bearden, *Chocolate: A Healthy Passion*, Prometheus Books, Amherst, N. Y., 2008, p. 144. For more on the "chemistry of love", see Michael R. Liebowitz, *The Chemistry of Love*, Little Brown, Boston, 1983, and for an overview of the biochemical mechanisms involved that relate to chocolate, see David Benton, "The Biology and Psychology of Chocolate Craving", in *Coffee, Tea, Chocolate, and the Brain*, ed. Astrid Nehlig, CRC Press, Boca Raton, FL, 2004, pp. 205–218.
27. W. J. Hurst and P. B. Toomey, "High performance liquid chromatographic determination of four biogenic amines in chocolate", *Analyst*, 1981, **106**, pp. 394–404; W. J. Hurst, R. A. Martin, B. L. Zoumas, *et al.*, "Biogenic amines in chocolate", *Nutrition Reports International*, 1982, **26**, pp. 1081–1086; and

W. J. Hurst, "A review of HPLC methods for the determination of selected biogenic amines in foods", *Journal of Liquid Chromatography*, 1990, **13**, pp. 1–23.
28. Katherine Khodorowsky and Hervé Robert, *The Little Book of Chocolate*, Flammarion, Luzon, France, 2001, p. 92.
29. S. Rössner, "Chocolate – divine food, fattening junk or nutritious supplementation?", *European Journal of Clinical Nutrition*, 1997, **51**, pp. 341–345, p. 343.
30. Linda K. Fuller, *Chocolate Fads, Folklore & Fantasies: 1,000 + Chunks of Chocolate Information*, Haworth Press, New York, 1994, p. 17.
31. Sophie D. Coe and Michael D. Coe, *The True History of Chocolate*, Thames and Hudson, London, 1996, p. 112.
32. Andrew Weil and Winifred Rosen, *From Chocolate to Morphine: Everything You Need to Know about Mind-Altering Drugs*, Revised edition, Houghton Mifflin, Boston, 2004, p. 48. Although a true gendered history of chocolate awaits an author, pharmacologist and ecofeminist scholar Cat Cox's *Chocolate Unwrapped: The Politics of Pleasure*, Women's Environmental Network, London, 1993, provides a remarkable starting point. Cultural anthropologist Emma Robertson adds another poignant yet startling view of women and chocolate in her *Chocolate, Women and Empire: A Social and Cultural History*, Manchester University Press, Manchester, 2010. Marcy Norton's *Sacred Gifts, A History of Profane Tobacco and Chocolate Pleasures in the Atlantic World*, Cornell University Press, Ithaca and London, 2008, illustrates the importance of chocolate in the lives of women in Mesoamerica and throughout the Atlantic World, especially regarding their marriageability, menstruation and social networking. In cultures where women secured their livelihood by becoming skilled in a desired craft, the ability to make good chocolate could raise a woman from the situation of a slave slated for death to a noblewoman. In this sense, chocolate literally became central to life or death.
33. Kristen Bruinsma and Douglas L. Taren, "Chocolate: Food or drug?", *Journal of the American Dietetic Association*, 1999, **99**, pp. 1249–1256.
34. Catherine S. Elliott, "Curing Irrationality with Chocolate Addiction", in *Chocolate: Food of the Gods*, ed. Alex Szogyi,

Greenwood Press for Hofstra University, Westport, CT, 1997, pp. 19–34, pp. 29–30.

35. Enrico Molinari and Edward Callus, "Psychological Drivers of Chocolate Consumption", in *Chocolate and Health*, ed. Rodolfo Paoletti, Andrea Poli, Ario Conti, *et al.*, Springer Verlag Italia, Milan, 2012, pp. 137–146.
36. Larry D. Reid, "Delicious or Addictive?", in *Chocolate, Fast Foods and Sweeteners: Consumption and Health,* Marlene R. Bishop, ed., Nova Science Publishers, New York, 2010, pp. 313–317.
37. Pennsylvania's Susquehanna University psychologist Debra A. Zellner chides the view of chocolate cravings merely in biophysiological terms, favoring instead conceptualising cravings as culturally defined. (Santiago Londoño Vélez, *The Virtues and Delights of Chocolate*, Compañía Nacional de Chocolates, Medellín, Colombia, 2003, pp. 108–109.)
38. Jeff Morgan, "Chocolate: A flavor and texture unlike any other", *American Journal of Clinical Nutrition*, 1994, **60 Supplement**, pp. 1065S–1067S focuses upon chocolate's "flavor and texture unlike any other". Many have claimed, with considerable evidence, that chocolate is the most commonly craved food. See, for instance, Marion M. Hetherington and Jennifer I. Macdiarmid, "'Chocolate addiction': A preliminary study of its description and its relationship to problem eating", *Appetite*, 1993, **21**, pp. 233–246, p. 233; David Benton, "The Biology and Psychology of Chocolate Craving", in *Coffee, Tea, Chocolate, and the Brain*, ed. Astrid Nehlig, CRC Press, Boca Raton, FL, 2004, pp. 205–218, p. 215; Enrico Molinari and Edward Callus, "Psychological Drivers of Chocolate Consumption", in *Chocolate and Health*, ed. Rodolfo Paoletti, Andrea Poli, Ario Conti, *et al.*, Springer Verlag Italia, Milan, 2012, pp. 137–146, p. 143.
39. Katherine Khodorowsky and Hervé Robert, *The Little Book of Chocolate*, Flammarion, Luzon, France, 2001, p. 26.
40. R. P. Hudson, "Polypharmacy in twentieth century America", *Clinical Pharmacology & Therapeutics*, 1968, **9**, pp. 2–10.
41. N. K. Hollenberg, "Chocolate: God's gift to mankind? Maybe!", 2005, http://www.earthtimes.org/article/news/3849.html, accessed 31 May 2011.

42. Expanding interest in the apparent correlations between the concentration of cacao within chocolate and the healthiness of the product has prompted many chocolate companies across the globe since the early 1990s to identify these percentages on product labels. Since the 1970s, individuals with lactose intolerance have been able to turn to chocolate-flavored soy milk. Though eaten in Saint John's Bread since antiquity, the beans of the carob tree have also become popular in recent years as a close-tasting alternative to chocolate for people concerned about cacao's high fat and alkaloid content (William Gervase Clarence-Smith, *Cocoa and Chocolate, 1765–1914*, Routledge, London and New York, 2000, p. 41).
43. L. Fernández-Murga, J. J. Tarín, M. A. García-Perez, *et al.*, "The impact of chocolate on cardiovascular health", *Maturitas*, 2011, **69**, pp. 312–321, p. 312.
44. B. H. Kean, "The blood pressure of the Cuna Indians", *American Journal of Tropical Medicine and Hygiene*, 1944, **24**, pp. 341–343, and N. K. Hollenberg, Gregorio Martinez, Marji McCullough, *et al.*, "Aging, acculturation, salt intake, and hypertension in the Kuna of Panama", *Hypertension*, 1997, **29**, pp. 171–176.
45. L. Djoussé, P. N. Hopkins, D. K. Arnett, *et al.*, "Chocolate consumption is inversely associated with calcified atherosclerotic plaque in the coronary arteries: The NHLBI Family Heart Study", *Clinical Nutrition*, 2011, **30**, pp. 38–43. Epub 2010 Jul 22, and L. Djoussé, P. N. Hopkins, K. E. North, *et al.*, "Chocolate consumption is inversely associated with prevalent coronary heart disease: The national heart, lung, and blood Institute Family Heart Study", *Clinical Nutrition*, 2011 Apr, **30**, pp. 182–187. Epub 2010 Sep 19.
46. B. Buijsse, E. J. Feskens, F. J. Kok, *et al.*, "Cocoa intake, blood pressure, and cardiovascular mortality: The Zutphen elderly study", *Archives of Internal Medicine*, 2006, **166**, pp. 411–417; B. Buijsse, C. Weikert, D. Drogan, *et al.*, "Chocolate consumption in relation to blood pressure and risk of cardiovascular disease in German adults", *European Heart Journal*, 2010, **31**, pp. 1616–1623. Epub 2010 Mar 30; Pamela J. Mink, Carolyn G. Scrafford, Leila M. Barraj, *et al.*, "Flavonoid intake and cardiovascular disease mortality: A prospective study in postmenopausal women", *American Journal of Clinical Nutrition*, 2007, **85**,

pp. 895–909; and E. Mostofsky, E. B. Levitan, A. Wolk, *et al.*, "Chocolate intake and incidence of heart failure: A population-based prospective study of middle-aged and elderly women", *Circulation. Heart Failure*, 2010, **3**, pp. 612–616. Epub 2010 Aug 16.

47. Curiously, the biomedical literature of recent decades seems to frequently substitute "cocoa" for "cacao" in its official terminology; thus, we follow this pattern of using "cocoa" within the phrasing of this section.
48. Francisco J. Pérez-Cano, Teresa Pérez-Berezo, Sara Ramos-Romero, *et al.*, "Is There an Anti-Inflammatory Potential Beyond the Antioxidant Power of Cocoa?", in *Chocolate, Fast Foods and Sweeteners: Consumption and Health*, ed. Marlene R. Bishop, Nova Science Publishers, New York, 2010, pp. 85–104. With increasing age, bodily cells, proteins and DNA become more noticeably damaged by oxidative processes that are largely mediated by the oxidative molecules' highly charged free radicals. Antioxidants help to neutralise this process, and the antioxidant power of specific foods and beverages is comparable by measuring their oxygen radical absorbance capacity (or ORAC).
49. Ki Won Lee, Nam Joo Kang, Min-Ho Oak, *et al.*, "Cocoa procyanidins inhibit expression and activation of MMP-2 in vascular smooth muscle cells by direct inhibition of MEK and MT1-MMP activities", *Cardiovascular Research*, 2008, **79**, pp. 34–41.
50. R. G. Baker, M. S. Hayden and S. Ghosh, "NF-KB, inflammation, and metabolic disease", *Cell Metabolism*, 2011, **13**, pp. 11–22.
51. Carlo Selmi, Tin K. Mao, Carl L. Keen, *et al.*, "The anti-inflammatory properties of cocoa flavanols", *Journal of Cardiovascular Pharmacology*, 2006, **47 Supplement 2**, pp. S163–S171.
52. Tankred Schewe, Hartmut Kühn and Helmut Sies, "Flavonoids of cocoa inhibit recombinant human 5-lipoxygenase", *JN: The Journal of Nutrition*, 2002, **132**, pp. 1825–1829.
53. A. Spadafranca, C. Martinez Conesa, S. Sirini, *et al.*, "Effect of dark chocolate on plasma epicatechin levels, DNA resistance to oxidative stress and total antioxidant activity in healthy subjects", *British Journal of Nutrition*, 2010, **103**, 1008–1014. Epub 2009 Nov 5.

54. Christian Heiss, Sarah Jahn, Melanie Taylor, *et al.*, "Improvement of endothelial function with dietary flavanols is associated with mobilization of circulating angiogenic cells in patients with coronary artery disease", *Journal of the American College of Cardiology*, 2010, **56**, pp. 218–224. For a review of the potential immunomodulator effects of chocolate, see Emma Ramiro-Puig and Margarida Castell, "Cocoa: Antioxidant and immunomodulator", *British Journal of Nutrition*, 2009, **101**, pp. 931–940.
55. Platelet activation, which participates in formation of atherosclerotic plaques, remains a target of study for comparing therapeutic agents. Dietrich Rein, Teresa G. Paglieroni, Ted Wun, *et al.*, "Cocoa inhibits platelet activation and function", *The American Journal of Clinical Nutrition*, 2000, **72**, pp. 30–35. See also B. Bordeaux, L. R. Yanek, T. F. Moy, *et al.*, "Casual chocolate consumption and inhibition of platelet function", *Preventive Cardiology*, 2007, **10**, pp. 175–180, for evidence of significantly decreased platelet activity following consumption of commercially available chocolate.
56. Debra A. Pearson, Teresa G. Paglieroni, Dietrich Rein, *et al.*, "The effects of flavanol-rich cocoa and aspirin on ex vivo platelet function", *Thrombosis Research*, 2002, **106**, pp. 191–197.
57. Michaël Hertog, G. L., Daan Kromhout, Christ Aravanis, *et al.*, "Flavonoid intake and long-term risk of coronary heart disease and cancer in the Seven Countries Study", *Archives of Internal Medicine*, 1995, **155**, pp. 381–386; Laura Yochum, Lawrence H. Kushi, Katie Meyer, *et al.*, "Dietary flavonoid intake and risk of cardiovascular disease in postmenopausal women", *American Journal of Epidemiology*, 1999, **149**, pp. 943–949; Paul Knekt, Jorma Kumpulainen, Ritva Järvinen, *et al.*, "Flavonoid intake and risk of chronic diseases", *American Journal of Clinical Nutrition*, 2002, **76**, pp. 560–568; and Pamela J. Mink, Carolyn G. Scrafford, Leila Barraj, *et al.*, "Flavonoid intake and cardiovascular disease mortality: A prospective study in postmenopausal women", *American Journal of Clinical Nutrition*, 2007, **85**, pp. 895–909.
58. Naomi D. L. Fisher and Norman K. Hollenberg, "Flavanols for cardiovascular health: The science behind the sweetness", *Journal of Hypertension*, 2005, **23**, pp. 1453–1459; Karen A.

Cooper, Jennifer L. Donovan, Andrew L. Waterhouse, *et al.*, "Cocoa and health: A decade of research", *British Journal of Nutrition*, 2008, **99**, pp. 1–11; Roberto Corti, Andreas J. Flammer, Norman K. Hollenberg, *et al.*, "Cocoa and cardiovascular health", *Circulation*, 2009, **119**, pp. 1433–1441; and Mark G. Shrime, Scott R. Bauer, Anna C. McDonald, *et al.*, "Flavonoid-rich cocoa consumption affects multiple cardiovascular risk factors in a meta-analysis of short-term studies", *The Journal of Nutrition*, 2011, **141**, pp. 1982–1988.

59. Dirk Taubert, R. Rosen, C. Lehmann, *et al.*, "Effects of low habitual cocoa intake on blood pressure and bioactive nitric oxide", *JAMA*, 2007, **298**, pp. 49–60.
60. S. T. Francis, K. Head, P. G. Morris, *et al.*, "The effect of flavanol-rich cocoa on the fMRI response to a cognitive task in healthy young people", *Journal of Cardiovascular Pharmacology*, 2006, **47**, pp. S215–S220.
61. Andreas J. Flammer, Frank Hermann, Isabella Sudano, *et al.*, "Dark chocolate improves coronary vasomotion and reduces platelet reactivity", *Circulation*, 2007, **116** , pp. 2376–2382.
62. Yumi Shinna, Nobusada Funabashi, Kwangho Lee, *et al.*, "Acute effect of oral flavonoid-rich dark chocolate intake on coronary circulation, as compared with non-flavonoid white chocolate, by transthoracic Doppler echocardiography in healthy adults", *International Journal of Cardiology*, 2009, **131**, pp. 424–429.
63. A. Buitrago-Lopez, J. Sanderson, L. Johnson, *et al.*, "Chocolate consumption and cardiometabolic disorders: Systematic review and meta-analysis", *BMJ*, 2011, **343**, d4488.
64. Mark G. Shrime, Scott R. Bauer, Anna C. McDonald, *et al.*, "Flavonoid-rich cocoa consumption affects multiple cardiovascular risk factors in a meta-analysis of short-term studies", *The Journal of Nutrition*, 2011, **141**, pp. 1982–1988.
65. Maximilian Schuier, Helmut Sies, Beate Illek, *et al.*, "Cocoa-related flavonoids inhibit CFTR-mediated chloride transport across T84 human colon epithelia", *Journal of Nutrition*, 2005, **135**, pp. 2320–2325.
66. Omar S. Usmani, Maria G. Belvisi, Hema J. Patel, *et al.*, "Theobromine inhibits sensory nerve activation and cough", *The FASEB Journal*, 2005, **19**, 231–233.

67. Monique Lacroix, "Polyphenols in Cocoa: Influence of Processes on Their Composition and Biological Activities", in *Chocolate, Fast Foods and Sweeteners: Consumption and Health*, ed. Marlene R. Bishop, Nova Science Publishers, New York, 2010, pp. 183–197, p. 191. For some years, the ability of polyphenols to activate tumor suppressor genes, thereby blocking the propagation of cancer, has been under investigation.
68. Carlo Selmi, Claudio A. Cocchi, Mario Lanfredini, *et al.*, "Chocolate at heart: The anti-inflammatory impact of cocoa flavanols", *Molecular Nutrition & Food Research*, 2008, **52**, pp. 1340–1348, p. 1345.
69. Beatrice A. Golomb, Sabrina Koperski and Halbert L. White, "Association between more frequent chocolate consumption and lower body mass index", *Archives of Internal Medicine*, 2012, **172**, pp. 519–521, p. 520.
70. Carlo Selmi, Claudio A. Cocchi, Mario Lanfredini, *et al.*, "Chocolate at heart: The anti-inflammatory impact of cocoa flavonols", *Molecular Nutrition & Food Research*, 2008, **52**, pp. 1340–1348. For other recent reviews, see Roberto Corti, Andreas J. Flammer, Normal K. Hollenberg, *et al.*, "Cocoa and cardiovascular health", *Circulation*, 2009, **119**, pp. 1433–1441 and Francesco Visioli, H. Bernaert, R. Corti, *et al.*, "Chocolate, lifestyle, and health", *Critical Reviews in Food Science and Nutrition*, 2009, **49**, pp. 299–312.
71. Karen A. Cooper, Jennifer L. Donovan, Andrew L. Waterhouse, *et al.*, "Cocoa and health: A decade of research", *British Journal of Nutrition*, 2008, **99**, pp. 1–11, p. 9.
72. M. Rusconi and A. Conti, "*Theobroma cacao L.*, the food of the Gods: A scientific approach beyond myths and claims", *Pharmacological Research*, 2010, **61**, pp. 5–13, p. 5.
73. See also the introductory chapter of that volume, Philip K. Wilson, "Chocolate as Medicine: A Changing Framework of Evidence Throughout History", in *Chocolate and Health*, eds. Rodolfo Paoletti, Andrea Poli, Ario Conti, and Francesco Visioli, Springer-Verlag Italia, Milan, 2012, pp. 1–15.

Epilogue: Prognosticating Chocolate's Future as Medicine

> The persons who habitually take chocolate are those who enjoy the most equable and constant health and are least liable to a multitude of illnesses which spoil the enjoyment of life.
>
> Jean Anthelme Brillat-Savarin, *Physiologie du Goût, ou Méditations de Gastronomie Transcendante* (1825), as cited in Robert Whymper (1912).

> The age of chocolate has arrived.
>
> Julie Pech, The Chocolate Therapist™: *A User's Guide to the Extraordinary Health Benefits of Chocolate* (2006)

A century ago, Robert Whymper, a prominent chemist in the chocolate industry noted that, "whether from the fact that cacao preparations do actually assist in maintaining good health, or for the reason that chocolate is the most pleasing confection to the palate, it is certain that the growth of the cacao industry in the last three centuries is little short of remarkable".[1] Perhaps, in retrospect, there has long been little difficulty in getting people to consume chocolate as a remedy for disease. Indeed, there has been little difficulty in getting people to consume chocolate for

Chocolate as Medicine: A Quest over the Centuries
Philip K. Wilson and W. Jeffrey Hurst

Published by the Royal Society of Chemistry, www.rsc.org

any reason. People from all walks of life find chocolate to be essential. As noted by the renowned 19th century Colombian author, Juan Francisco Ortíz, chocolate is

> good for the sick and the elderly; children and old people drink it doggedly. The traveler in Nueva Granda always takes some bars of chocolate with him in his baggage. Friars and monks think of chocolate when they are singing vespers, workers take great delight in drinking a cup from time to time; and chocolate is the first thing that sufferers from migraine, constipation, toothache, every kind of illness think of.

And then he poignantly, elegantly and concisely concludes, chocolate is "the comforter of the afflicted".[2]

Since the introduction of this New World wonder to the Old World, chocolate has never fallen out of fashion. Long before identifying abundant qualities of phenylethylamine (PEA) in chocolate, the soothing, euphoric effects of consuming chocolate were widely known. Indeed, chocolate markedly exemplifies *Botany of Desire* author Michael Pollan's depiction of ways in which plants have, through a symbiotic relationship, shaped humanity in terms of satisfying and gratifying human desire. Part of chocolate's success also derived from it being a "well-rounded food, containing glucose, lipids, and proteins. It also provides significant quantities of important minerals, such as potassium and calcium. Of course, such additives as milk and sugar bring other qualities and calories".[3]

In recent decades, chocolate has become featured as the substance of significant genetic research. In the 1990s, the International Clone Trial was established to expand the worldwide distribution of cacao genotypes with stable and disease-resistant traits. Hybrid trials were also undertaken in order to create new cacao varieties with genotypes that would allow for an increasingly sustainable chocolate product at lower costs.[4] Two decades later, an international team, including Professors Mark Guiltinan, Siela Maximova and other researchers from horticulture, plant biology, biochemistry and forest resources at Penn State University (Pennsylvania being the largest chocolate products producing state in the United States) have sequenced the genome of the *Criollo* variety of the Chocolate Tree – the variety generally used in the production of the world's finest chocolate.[5] As such, *Theobroma cacao* was the first early domesticated tropical fruit tree to have its genome sequenced. This information also allowed for the identification of gene families on the cacao chromosomes from which scientists hope to genetically enhance the ability of future Chocolate Tree crops to resist the fungal diseases and insects that target these trees, thereby increasing their product yield. In addition, altering the genome might allow for the production of "even healthier chocolate".[6] Such matters may take some time, for as science has taught us, nature frequently resists the attempts or at least the speed of the attempts to gain human control over it.[7] The USDA Agricultural Research Service has, in cooperation with a number of other companies and institutions including Mars, Inc., concentrated on cacao genome sequencing regarding other varieties of *Theobroma cacao*, and they are largely responsible for creating and maintaining the Cacao Genome Database.

One other initiative of the Pennsylvania State University system is the work undertaken by its Cocoa, Chocolate and Confectionery Research Group. There, Food Science Professors including Ramaswamy C. Anantheswaran, John Coupland, John E. Hayes, Joshua D. Lambert and Gregory R. Ziegler conduct collaborative and creative research on cocoa, chocolate and confectionery science and technology that serves confectionery manufacturers and consumers.

On the matter of chocolate and health, genetic research from Japan strongly suggests that it is the polyphenols in cacao that bind to specific genes in the liver and intestine, thereby activating other genes to produce HDL as well as to suppress those genes responsible for producing LDL. The ability of cocoa-derived polyphenols to modulate cell signaling and gene expression is one newly developing focus toward a better understanding of the biomechanics underlying chocolate and health.

Chocolate has also long been part of what has been variously called alternative medicine, complementary medicine or more recently, integrative holistic medicine. In general, many of these practices have withstood the test of time, at least within certain cultural contexts. Though skepticism continues to surround at least some of these medical pursuits, they are not a fad. Today, integrative medicine practitioners focus upon each patient as a unique individual, noting in particular variations in how particular patients may respond to diseases or to treatments. Their chief task is to integrate what works for that individual patient, at times varying treatments based upon specific cultural expectations. This focus upon the individual also underlies what is gradually becoming recognised as the practise of personalised medicine.

Among these approaches to healing, chocolate has been one of the natural substances within the armamentarium of practitioners of alternative, complementary and now integrative medicine. As we have seen, cacao has been used to treat or prevent a wide variety of ailments throughout history. Within the past few decades, computer-accessible databases of information have been created whereby both healers and health-seeking patients can easily obtain information on the various preparations of cacao or cacao-containing products deemed useful in treating particular ailments. Of course, just what individual or organisation deems them to be useful at times depends upon the database consulted. Among the natural product internet sites that medical authorities most regularly consult with confidence are Dr James Duke's Phytochemical and Ethnobotanical Databases, the Natural Medicines Comprehensive Database, and the website of the National Institutes of Health National Center for Complementary and Alternative Medicine. Chocolate's reputed benefits are discussed within these and other commonly used sites.

Chocolate has been used as both food and medicine in China. Antoine Gallais described the high Chinese consumption of cacao (which they imported in the form of a paste and added their own spices) in his *Monographie du Cacao, ou Manual de l'Amateur de Chocolat* (1827). Alexandre Dumas featured Chinese chocolate use in his *Grand Dictionnaire de Cuisine* (1873). More recently, China held its first international exposition of fine chocolate – *Salon de Chocolat* – in Beijing in 2004. Perhaps an increasing use of chocolate in Traditional Chinese Medicine remedies will accommodate the expanding market of chocolate in China.[8]

Relatedly, many natural product sites describe chocolate as a "functional food" that modulates health. The Institute of Medicine's Food and Nutrition Board defines such foods as those that "may provide a health benefit beyond the traditional nutrients it contains".[9] Though related, "functional food" is conceptually distinct from earlier references to chocolate as a "protective food", a term that health crusader Bernarr MacFadden used nearly a century ago to convey the sense that eating chocolate served to "protect" us by making up for deficiencies found within other foods of our diet.[10] In 1925, J.R. Snavely concurred with such a view, further arguing that chocolate was nutritious in that it supplied "heat and energy and some minerals for growth", though he denied earlier claims that chocolate was a "complete food" as it could not "in itself supply all the needs of the human body".[11] Functional foods, on the other hand, provide benefit beyond their mere nutritional value. Attempts to better determine public attitudes and behaviors related to the perception of chocolate as a functional food are underway.[12] We are likely to see chocolate increasingly branded as a functional food and/or nutraceutical for an expanding variety of health-related benefits, particularly as the connections between chocolate's ability to enhance nutrition while minimising disease becomes more paramount in overcoming the obesity epidemic which a growing number of people in many parts of the world are facing.

Another recent and exciting potential health benefit of chocolate has come from its use in the care of dementia patients. As part of the comfort-centered care offered in facilities such as the Beatitudes Nursing Home in Phoenix, Arizona, chocolate is regularly administered to patients, and its effects are noted in the nursing charts. If, as behavioral research suggests, creating facilities that are filled with positive emotional environments for people with significant dementia can diminish patient's inexplicable yet measurable signs of distress (and other behavioral problems such as late afternoon "sundowning"), then perhaps more positive outcomes are achievable simply through chocolate.[13]

Many people become less stressed when they experience something that once produced great joy in their lives. Chocolate has been shown to be strongly "linked to memories of childhood, the maternal instinct, and affection". Overall, it mentally triggers feelings of – or at least associations with – "warmth and protection, reminding us of situations that are pleasant and familiar".[14] Thus, if chocolate had once been a favored comfort food, why not offer it as needed to people with dementia who, as a particular group of humans, need added human comfort? Additionally, as Beck and Damkjaer have suggested, chocolate may help curb the weight loss and decrease in the body mass index commonly seen among those living in extended care facilities including nursing homes.[15] In terms of nutritional intervention, chocolate may be a quality of life extender as it boosts the nutritional, physical and mental health of the growing number of elderly individuals who are admitted to long-term care facilities.

Might "chocolate therapy" become a future medical or psychology specialty? Chloé Doutre-Roussel, once a chocolate buyer for London's Fortnum & Mason and now author of the highly acclaimed, *The Chocolate Connoisseur* (2006), has widely shared her vision of "real, good-quality, high cacao-content chocolate" being "recommended by doctors for its pleasurable, anti-depressant qualities, just as they might recommend a glass of red wine to fight heart disease". At the very least, she continues, this would "make people happier and stronger" in the face of illness.[16] Happiness may also be at the heart of the Van Leer Chocolate Company employing a "chocolate doctor" who "makes house calls or does telephone consultations" from its Jersey City, New Jersey office. "Healers" across the globe are conducting chocolate therapy sessions. Even when not at one of those sessions, a more globally accessible individual, Julie Pech, advertises her services as "The Chocolate Therapist™" on the web at www.thechocolatetherapist.com. Chocolate's reputed health benefits are also being capitalised upon by cookbook authors and gourmands, such as Janette Marshall's *The Here's Health Alternative Chocolate Book* (1986), Rowan Jacobsen's *Chocolate Unwrapped: The Surprising Health Benefits of America's Favorite Passion* (2003) and Victoria Laine's *Health By Chocolate: Radical New Recipes & Nutritional Know-How* (2008).

As a reminder of the persistent availability of chocolate in spa towns, one need only look at the website ads that aim to persuade us to "Melt into the ultimate luxury at the decadent Spa At The Hotel Hershey, affectionately known as The Chocolate Spa". There, one can obtain, in addition to traditional massages, "signature treatments like our Whipped Cocoa Bath and Cocoa Facial Experience". Additional services include the Coconut Chocolate Truffle Immersion Body Treatment, the Chocolate & Roses Decadence Body Treatment, the Chocolate Bean Polish, the Chocolate Sugar Scrub, the Cocoa Massage, Chocolate Hydrotherapy or the tantalizing Chocolate Fondue Wrap. All such offerings seemingly provide further support for Hershey's reputation as being the "biggest success in the world of chocolate".[17]

In the United States, it was Mars, Inc., that in recent years was the first major chocolate company to redirect cacao research towards identifying its health potential. They altered their cacao bean processing to yield a product with a high flavanol and procyanidin levels, trademarking this bean as "cocoapro". Their health-themed chocolate product Cocoa Via, a granola-type bar fortified with antioxidants, did not fully mask the naturally bitter-tasting flavanols, and thus did not strike the consumer palate as desired. Still, the race for researching, developing and patenting a wide variety of cacao-rich products that have promising potential in terms of their health benefits has begun, both with the industrial giants like the Hershey Company as well as with small artisan ventures like Theo Chocolates of Freemont, Washington, under the health-conscious guidance of Chief Operating Officer, Dr Andy McShea, aka "Doc Choc". In addition to focusing on the healthiness of the product, Hershey's has also recently undertaken specific measures to improve the healthy surrounds of cacao farmers, particularly in West Africa where currently they are obtaining their largest supplies of cacao.

A Nestlé Company publication of many decades ago, *The History of Chocolate and Cocoa*, identified chocolate's "coming-of-age" in the world in terms of four important turning points.

1. introducing cocoa powder;
2. reducing excise duties;
3. improving transportation from plantation to factory; and
4. inventing and improving the manufacture of eating chocolate.[18]

In the near future, might "Substantiating its medical and health benefits" be the fifth turning point in chocolate's odyssey?

Is "prescription strength chocolate" just around the corner? Several potential chocolate company–pharmaceutical industry partnerships are in the talking stages of joint ventures to develop chocolate into a prescription drug form. But taking chocolate in pill form, something akin to the purified cacao tablets already available, may well distort the pleasure-filled aspect so critical to chocolate consumption. Perhaps prescription strength chocolate would work best as an additive to that which is already being ingested as a matter of habit.

"Death by Chocolate" is a phrase that many, including Sandra Boynton in her clever *Chocolate: The Consuming Passion* (1982), refer to as a specially craved dessert, but evidence of all types seems to indicate that, regardless of how you look at it, chocolate is very good for the living. Even in popular fiction, J.K. Rowling depicted chocolate as Harry Potter's only life-saving first aid against the soul-consuming Dementors. Other fictitious characters seemingly become chocolate itself. For instance, Robert Kimell Smith's 1972 medical casebook for children recounts how Henry Green, whom some people believed "wasn't really born, but was hatched fully grown, from a chocolate bean", began to smell and taste like chocolate after he broke out with a "chocolate fever".[19] Young Henry might just be the exception to Brillat-Savarin's edict that chocolate is actually useful to "calm a fever" and to "restore" bodies to complete health. Building upon the wisdom of Forrest Gump's quip that life is like a box of chocolates, various forms of evidence suggest that life may certainly benefit from at least an occasional box of chocolates.

Though this book has admittedly provided only snippets from the vast history of chocolate's luscious timeline, the repeated attempts over the past 400 plus years to gather evidence for chocolate's therapeutic capacity lead us to surmise that this quest itself, as well as the substance being studied, has something approaching a habit-forming nature to it.

REFERENCES

1. Robert Whymper, *Cocoa and Chocolate: Their Chemistry and Manufacture*, J. & A. Churchill, London, 1912, p. ix.
2. Juan Francisco Ortíz, as cited in Santiago Londoño Vélez, *The Virtues and Delights of Chocolate*, Compañía Nacional de Chocolates, Medellín, Colombia, 2003, p. 58.
3. Murdo J. MacLeod, "Cacao", in *The Cambridge World History of Food*, eds. Kenneth F. Kiple and Kriemhild Coneè Ornelas, Cambridge University Press, Cambridge, UK and New York, 2000, Vol. 1, p. 640.
4. A. B. Eskes and Y. Efron, ed., *Global Approaches to Cocoa Germplasm Utilization and Conservation*, CFC, Amsterdam, The Netherlands, 2006, preface.

5. X. Argout, J. Salse, J. M. Aury, *et al.*, "The Genome of *Theobroma cacao*", *Nature Genetics*, 2011, **43**, pp. 101–108 and Mark J. Guiltinan, "Genomics of *Theobroma cacao*, 'the Food of the Gods,'" in *Genomics of Tropical Crop Plants*, eds. P. H. Moore and R. Ming, Springer, New York, 2008, pp. 145–170.
6. A'ndrea Elyse Messer, "Cocoa (data) crunch", *Research Penn State*, 2011, **32**, p. 21.
7. For more on the variability and diversity of cacao in the modern era, see B. G. D. Bartley, *The Genetic Diversity of Cacao and its Utilization*, CABI Pub., Cambridge, MA, 2005.
8. Lawrence L. Allen, *Chocolate Fortunes: The Battle for the Hearts, Minds, and Wallets of Chinese Consumers*, American Management Association, New York, 2010.
9. Deanna Pucciarelli and James Barrett, "Twenty-First Century Attitudes and Behaviors Regarding the Medicinal Use of Chocolate", in *Chocolate: History, Culture, and Heritage*, eds. Louis Evan Grivetti and Howard Yana Shapiro, John Wiley & Sons, Hoboken, N. J., 2009, pp. 651–666, pp. 653–654. For earlier work on chocolate as functional food, see Andrea Borchers, Carl L. Keen, Sandra M. Hannum, *et al.*, "Cocoa and chocolate: Composition, bioavailability, and health implications", *Journal of Medicinal Food*, 2000, **3**, pp. 77–105.
10. Bernarr Macfadden, *Hershey, The Chocolate Town*, Chocolate Company, Hershey, PA, ca. 1923.
11. J. R. Snavely, "The nutritive value of chocolate and cocoa", *The Confectioners' Journal*, 1925, **51**, p. 96.
12. Deanna Pucciarelli and James Barrett, "Twenty-First Century Attitudes and Behaviors Regarding the Medicinal Use of Chocolate", in *Chocolate: History, Culture, and Heritage*, eds. Louis Evan Grivetti and Howard Yana Shapiro, John Wiley & Sons, Hoboken, N. J., 2009, pp. 653–666.
13. Pam Belluck, "Giving Alzheimer's Patients Their Way, Even Chocolate", *The New York Times*, December 31, 2010, http://www.nytimes.com/2011/01/01/health/01care.html?_r=1&pagewanted=print, accessed 14 June 2011. William Alex McIntosh, Karen S. Kubena and Wendall A. Landmann, "Chocolate and Loneliness among the Elderly", in *Chocolate: Food of the Gods*, ed. Alex Szogyi, Greenwood Press for Hofstra University, Westport, CT, 1997, pp. 3–10 found that

chocolate benefited the general aging population in terms of stress relief.

14. Enrico Molinari and Edward Callus, "Psychological Drivers of Chocolate Consumption", in *Chocolate and Health*, eds. Rodolfo Paoletti, Andrea Poli, Ario Conti, *et al.*, Springer Verlag Italia, Milan, 2012, pp. 137–146, p. 137.
15. Anne Marie Beck and Karin Damkjoer, "Chocolate: A Significant Part of Nutrition Intervention among Elderly Nursing Home Residents", in *Chocolate, Fast Foods and Sweeteners: Consumption and Health*, ed. Marlene R. Bishop, Nova Science Publishers, New York, 2010, pp. 252–253.
16. Chloé Doutre-Roussel, *The Chocolate Connoisseur*, Penguin, New York, 2006, p. 14.
17. David Lebovitz, *The Great Book of Chocolate: The Chocolate Lover's Guide, with Recipes*, Ten Speed Press, Berkeley, 2004, p. 16.
18. Nestlé Company, *The History of Chocolate and Cocoa* [leaflet], n.d, p. 3.
19. Smith's *Chocolate Fever*, as described by Linda K. Fuller, *Chocolate Fads, Folklore & Fantasies: 1,000+ Chunks of Chocolate Information*, Haworth Press, New York, 1994, p. 21.

APPENDIX 1

Disorders and Diseases that Chocolate (Cacao) Products have Reputedly Improved throughout History[1]

Agitation (Reduces)
Anemia (Improves)
Angina/Heart Pain (Reduces)
Apathy (Diminishes)
Aphrodisiac Properties (Enhances)
Appetite (Improves)
Asthma (Diminishes)
Belching (Controls/Diminishes)
Bleeding (Staunches)
Blood (Generates)
Body (Fortifies/Invigorates/Nourishes/Refreshes/Repairs)
Brain (Strengthens)
Breath (Sweetens/Reduces Shortness of)
Bronchitis (Diminishes)
Burns (Soothes)
Cancer (Diminishes)

[1]Based upon the definitive listing prepared as Table 2 by Teresa Dillinger, Patricia Barriga, Sylvia Escárcega, *et al.*, "Food of the Gods: Cure for humanity? A cultural history of the medicinal and ritual use of chocolate", *JN: The Journal of Nutrition*, 2000, **130 Supplement**, pp. 2057S–2072S, pp. 2061S–2063S

Chocolate as Medicine: A Quest over the Centuries
Philip K. Wilson and W. Jeffrey Hurst

Published by the Royal Society of Chemistry, www.rsc.org

Catarrh (Diminishes)
Chest Ailments (Diminishes)
Colds (Diminishes)
Colic (Diminishes)
Conception (Improves Probability of)
Consumption/Tuberculosis (Diminishes)
Cough (Diminishes)
Countenance (Preserves)
Cuts (Soothes/Disinfects)
Debilitation (Improves)
Diarrhea/Dysentery/Griping of Guts/Dyspepsia (Diminishes)
Digestion (Improves/Promotes)
Disposition (Improves/Consoles)
Distempers (Diminishes)
Emaciation/Wasting (Diminishes)
Energy (Improves)
Exhaustion (Relieves)
Exercise (Nourishes Body after)
Fainting (Relieves)
Fatigue (Diminishes)
Female Complaints (Diminishes)
Fevers (Reduces/Relieves)
Flatulence/Wind (Controls/Dissipates/Reduces)
Gout (Diminishes)
Green Sickness/Chlorosis (Diminishes)
Gums (Strengthens)
Hair (Delays Growth of Whiteness)
Hangover (Reduces Effects of)
Hemorrhoids/Piles (Reduces)
Health (Preserves)
Heart (Strengthens)
Heart Palpitations (Relieves)
Hoarseness (Relieves)
Hypochondria (Diminishes)
Indigestion (Diminishes)
Infection (Reduces)
Inflammation (Reduces)
Itch (Diminishes)
Jaundice (Diminishes)
Kidney Complaints (Diminishes)

Kidney Stone (Expels/Cures)
Labor/Delivery/Childbirth (Facilitates)
Lactation (Promotes)
Leucorrhea/The "Whites" (Diminishes)
Life (Improves)
Limbs (Strengthens)
Lips, Chapped or Cracked (Soothes)
Liver Complaints (Diminishes)
Longevity (Lengthens/Prolongs)
Lung Complaints (Diminishes)
Menstruation (Provokes/Increases)
Morality (Improves)
Mouth, Burning (Relieves)
Nervous Disorders (Calms/Diminishes)
Nipples, Cracked (Soothes)
Nutrition (Improves)
Obstructions (Opens/Reduces)
Pain (Eases/Reduces)
Poison (Counters/Expels/Antidote)
Pregnancy (Nourishes Embryo)
Rectal Bleeding (Diminishes)
Rheumatism (Diminishes)
Scurvy (Diminishes)
Seizures (Reduces)
Sexual Appetite/Desire/Pleasure (Arouses/Increases)
Skin (Clears/Lubricates/Softens)
Skin Eruptions (Diminishes)
Sleep (Encourages and Prevents)
Spirit (Invigorates)
Spleen (Deadens)
St. Anthony's Fire/Erysipelas (Diminishes)
Strength (Recovers/Repairs)
Sweat (Provokes/Increases)
Syphilis (Diminishes)
Teeth (Cleans)
Thirst (Quenches)
Throat Inflammation (Reduces)
Timidity (Diminishes)
Toothache (Eases/Reduces)
Tumors/Pustules (Reduces)

Ulcers (Reduces)
Urine Flow (Provokes/Increases)
Vaginal Irritation (Reduces)
Virility (Increases)
Violence (Diminishes)
Vomiting (Controls/Reduces)
Warmth (Increases)
Weakness (Relieves)
Weight (Increases)
Worms (Cures)
Wounds (Soothes/Disinfects)

APPENDIX 2

18th-Century General Recipe for "Health Chocolate"[1]

In order to prepare and make chocolate paste, the first step is to toast the cacao. Place a small amount of river sand, which is white, fine and has been sifted, to make it even finer, in an iron toaster in the form of a pan (that in Madrid this is called a *paila*), and put [it] over a fire. Heat the sand and stir it with a wooden stick so that it heats evenly, then add the cacao. Continue to stir it in with the sand until the heat penetrates the shell, which separates it from the bean without burning it. It is better to toast it this way instead of without the sand for the following reasons: First, the chocolate becomes more intense in color; second, it separates from the shell without burning the bean; and third and most importantly, the fatty, oily and volatile parts of the cacao are not dissipated. Therefore, it keeps all of its medicinal and stomachic qualities, which will be described below.

Once the worker or miller knows that it is well-toasted, he puts it in a sack or a bag and leaves it until the next day when he pours it

[1]Antonio Lavedan, *Tratado de Los Usos, Abusos, Propiendades y Virtudes del Tabaco, Café, Te y Chocolate* (Madrid: Imprenta Real, 1796), Chapter IV. The natural philosopher Francis Willughby (also Willoughby) recounted much more briefly a common Spanish recipe for chocolate preparation used in the 1660s in "A Relation of a Voyage Made Through a Great Part of Spain, *etc*", that appeared in the work of his European travel companion, John Ray's 1673 *Observations … Made in a Journey Through Part of the Low Countries, etc.*, as cited by Head, Brandon, *The Food of the Gods: A Popular Account of Cocoa*, R. Brimley Johnson, London, 1903, p. 58.

Chocolate as Medicine: A Quest over the Centuries
Philip K. Wilson and W. Jeffrey Hurst

Published by the Royal Society of Chemistry, www.rsc.org

through a sieve in order to separate the sand from the cacao. Then it is hulled and cleaned from its shell as follows: it is placed on a quadrilateral, somewhat arched rock with a thickness of three fingers. A small brazier is placed underneath the rock with a little bit of carbon fire. The miller smashes and grinds the cacao with another round cylindrical rock and works it into a roll, which he takes in his hands by either end, thereby making a paste which is then mixed with one-half or third parts sugar. Once this mixture is made, it is ground again. Then, while still hot it is placed in molds of tin or wood, according to individual custom. Sometimes it is placed on paper and made into rolls or blocks, where it is fixed and becomes solid quickly. Others mash it and make it into a paste called ground chocolate.

Prepared in this manner, it is called "Health Chocolate". Some people claim that it is good to mix in a small amount of vanilla in order to facilitate digestion because of its stomachic and tonic qualities.

When you want a chocolate that further delights the senses, you add a very fine powder made from vanilla and cinnamon. This is added after mixing the sugar, and then it is ground a third time lightly, in order to make a good mixture. It is claimed that with these aromas it is digested better. Some people who like aromas add a little essence of amber.

Some chocolate makers add pepper and ginger, and wise people should be careful to not take chocolate without first knowing its composition. In Spain, people do not usually add either of these two spices, but in other countries, they tend to use one or another, although this is not very common because it is so disagreeable.

The way to make and drink chocolate so that it has all of its properties, is the following: First, the squares or rolls of chocolate are cut into pieces, and placed in the chocolate pot with cold water, not hot. They are placed over a slow fire and stirred well with the chocolate beater until well dissolved. Chocolate should never be heated at a high temperature because it does not need to cook or boil when it is well dissolved. This would cause it to coagulate, and the oily or fatty part separates when it boils. Those with weak stomachs cannot digest it in this form, and therefore, it does not have the desired effects. It is even worse when already dissolved chocolate is kept and later boiled again when someone wants to drink it. In order to fully enjoy the benefits of chocolate, it should be drunk immediately after it is dissolved and stirred. Others reduce the squares of chocolate to powder and add it to boiling water in the chocolate pot. They dissolve it, stirring it well with the chocolate beater and immediately pour it into the serving cups. This is the best way to make chocolate because there is no time for the oil to separate.

APPENDIX 3

Partial List of Chemical Compounds Found in Cacao

Acetic acid, aesculetin, alanine, alkaloids, alpha-sitosterol, alpha-theosterol, amyl-acetate, amyl-alcohol, amyl-butyrate, amylase, apigenin-7-o-glucoside, arabinose, arachidic acid, arginine, ascorbic acid, ascorbic-acid-oxidase, aspariginase, beta-carotene, beta-sitosterol, beta-theosterol, biotin, caffeic acid, caffeine, calcium, campesterol, catalase, catechins, catechol, cellulase, cellulose, chlorogenic acid, chrysoeriol-7-o-glucoside, citric acid, coumarin, cyanidin, cyanidin-3-beta-l-arabinoside, cyanidin-3-galactoside, cyanidin-glycoside, cycloartanol, d-galactose, decarboxylase, dextrinase, diacetyl, dopamine, epigallocatechin, ergosterol, ferulic acid, formic acid, fructose, furfurol, galacturonic acid, gallocatechin, gentisic acid, glucose, glutamic acid, glycerin, glycerophosphatase, glycine, glycolic acid, glycosidase, haematin, histidine, i-butyric acid, idaein, invertase, isobutylacetate, isoleucine, isopropyl-acetate, isovitexin, kaempferol, l-epicatechin, leucine, leucocyanidins, linalool, linoleic acid, lipase, luteolin, luteolin-7-o-glucoside, lysine, lysophosphatidyl-choline, maleic acid, mannan, manninotriose, mannose, melibiose, mesoinositol, methylheptenone, n-butylacetate, n-nonacosane, niacin, nicotinamide, nicotinic- acid, nitrogen, nonanoic acid, o-hydroxyphenylacetic acid, octoic acid, oleic acid, oleo-dipalmatin, oleopalmitostearin, oxalic acid, p-anisic acid, p-coumaric acid, p-coumarylquinic acid, p-hydroxybenzoic acid, p-hydroxyphenylacetic acid,

Chocolate as Medicine: A Quest over the Centuries
Philip K. Wilson and W. Jeffrey Hurst

Published by the Royal Society of Chemistry, www.rsc.org

palmitic acid, palmitodiolen, pantothenic acid, pectin, pentose, peroxidase, phenylacetic acid, phenylalanine, phlobaphene, phosphatidyl-choline, phosphatidyl- ethanolamine, phosphatidyl-inositol, phospholipids, phosphorus, phytase, planteose, poly-galacturonate, polyphenol-oxidase, polyphenols, proline, propionic acid, propyl-acetate, protocatechuic acid, purine, pyridoxine, quercetin, quercetin-3-o-galactoside, quercetin-3-o-glucoside, quercitrin, raffinase, raffinose, reductase, rhamnose, riboflavin, rutin, rutoside, saccharose, salsolinol, serine, sinapic acid, stachyose, stearic acid, stearodiolein, stigmasterol, sucrose, syringic acid, tannins, tartaric acid, theobromine, theophylline, thiamin, threonine, trigonelline, tyramine, tyrosine, valerianic acid, valine, vanillic acid, verbascose, verbascotetrose, vitexin.

APPENDIX 4

Captain James Wadsworth's Poetical Introduction to his 1652 translation of Dr Antonio Colmenero de Ledesma's *Chocolate; or, An Indian Drinke. By the Wise and Moderate Use whereof, Health is Preserved, Sicknesse Diverted, and Cured, Especially the Plague of the Guts; Vulgarly called The New Disease; Fluxes, Consumptions, & Coughs of the Lungs, with Sundry other Desperate Diseases. By it also, Conception is Caused, the Birth Hastened and Facilitated, Beauty Gain'd and Continued.*

Chocolate as Medicine: A Quest over the Centuries
Philip K. Wilson and W. Jeffrey Hurst

Published by the Royal Society of Chemistry, www.rsc.org

To every Individuall Man, and Woman, Learn'd, or Unlearn'd, Honest, or Dishonest: In the Due Praise of Divine Chocolate.

Doctors lay by your Irksome Books
And all ye Petty-Fogging Rookes
Leave Quacking; and Enucleate
The vertues of our Chocolate.

Let th' Universall Medicine
(Made up of Dead-mens Bones and Skin,)
Be henceforth Illegitimate,
And yield to Soveraigne-Chocolate.

Let Bawdy-Baths be us'd no more;
Nor Smoaky-Stoves but by the whore
Of Babilon: since Happy-Fate
Hath Blessed us with Chocolate.

Let old Punctæus Greaze his shooes
With his Mock-Balsome: and Abuse
No more the World: But Meditate
The Excellence of Chocolate.

Let Doctor Trigg (who so Excells)
No longer Trudge to Westwood-Wells:
For though that water Expurgate,
'Tis but the Dreggs of Chocolate.

Let all the Paracelsian Crew
Who can Extract Christian from Jew;
Or out of Monarchy, A State,
Breake all their Stills for Chocolate.

Tell us no more of Weapon-Salve,
But rather Doome us to a Grave:
For sure our wounds will Ulcerate,
Unlesse they're wash'd with Chocolate.

The Thriving Saint, who will not come
Within a Sack-Shop's Bowzing-Roome
(His Spirit to Exhilerate)
Drinkes Bowles (at home) of Chocolate.

His Spouse when she (Brimfull of Sense)
Doth want her due Benevolence,
And Babes of Grace would Propagate,
Is alwayes Sipping Chocolate.

The Roaring-Crew of Gallant-Ones
Whose Marrow Rotts within their Bones:
Their Bodyes quickly Regulate,
If once but Sous'd in Chocolate.

Young Heires that have more Land then Wit,
When once they doe but Tast of it,
Will rather spend their whole Estate,
Then weaned be from Chocolate.

The Nut-Browne-Lasses of the Land
Whom Nature vayl'd in Face and Hand,
Are quickly Beauties of High-Rate,
By one small Draught of Chocolate.

Besides, it saves the Moneys lost
Each day in Patches, which did cost
Them deare, untill of Late
They found this Heavenly Chocolate.

Nor need the Women longer grieve
Who spend their Oyle, yet not conceive,
For 'tis a Helpe-Immediate,
If such but Lick of Chocolate.

Consumptions too (be well assur'd)
Are no lesse soone then soundly cur'd:
(Excepting such as doe Relate
Unto the Purse) by Chocolate.

Nay more: It's vertue is so much,
That if a Lady get a Touch,
Her griefe it will Extenuate,
If she but smell of Chocolate.

The Feeble-Man, whom Nature Tyes
To doe his Mistresse's Drudgeries;
O how it will his minde Elate,
If shee allow him Chocolate!

'Twill make Old women Young and Fresh;
Create New-Motions of the Flesh,
And cause them long for you know what,
If they but Tast of Chocolate.

There's ne're a Common Counsell-Man,
Whose Life would Reach unto a Span,
Should he not Well-Affect the State,
And First and Last Drinke Chocolate.

Nor e're a Citizen's Chast wife,
That ever shall prolong her Life,
(Whilst open stands Her Posterne-Gate)
Unlesse she drinke of Chocolate.

Nor dost the Levite any Harme,
It keepeth his Devotion warme,
And eke the Hayre upon his Pate,
So long as he drinkes Chocolate.

Both High and Low, both Rich and Poore
My Lord, my Lady, and his ——
With all the Folkes at Billingsgate,
Bow, Bow your Hamms to Chocolate.

Bibliography

Aaron, Shara, and Monica Bearden, *Chocolate: A Healthy Passion*, Prometheus Books, Amherst, N. Y., 2008.
Allen, Lawrence L., *Chocolate Fortunes: The Battle for the Hearts, Minds, and Wallets of Chinese Consumers*, American Management Association, New York, 2010.
Amado, Jorge, *Cacáu*, Ariel, Rio de Janeiro, 1933.
Arbuthnot, John, *An Essay Concerning the Nature of Ailments*, 2nd edition, J. Tonson, London, 1732.
Argout, X., J. Salse, J. M. Aury, *et al.*, "The genome of *Theobroma cacao*", *Nature Genetics*, 2011, **43**, pp. 101–108.
Bailleux, Nathalie, Hervé Bizeul, John Feltwell, *et al.*, *The Book of Chocolate*, Flammarion, Paris and New York, 1996.
Bainbridge, J. C., and S. H. Davies, "Essential oil of cocoa", *Journal of the Chemical Society, Transactions*, 1912, **101**, pp. 2209–2221.
Baker, R. G., M. S. Hayden, and S. Ghosh, "NF-KB, inflammation, and metabolic disease", *Cell Metabolism*, 2011, **13**, pp. 11–22.
Bancroft, Hubert Howe, *The Native Races of the Pacific States of North America*, Longmans, Green & Co., London, 1875.
Barthel, Diane, "Modernism and marketing: The chocolate box revisited", *Theory, Culture & Society*, 1989, **6**, pp. 429–438.
Bartley, B. G. D., *The Genetic Diversity of Cacao and its Utilization*, CABI Pub., Cambridge, MA, 2005.
Beale, J. F., Jr., "Cocoa of to-day and yesterday", *Confectioners' Journal*, 1906, **32**, pp. 84–85.

Chocolate as Medicine: A Quest over the Centuries
Philip K. Wilson and W. Jeffrey Hurst

Published by the Royal Society of Chemistry, www.rsc.org

Beck, Anne Marie, and Karin Damkjoer, "Chocolate: A Significant Part of Nutrition Intervention among Elderly Nursing Home Residents", in *Chocolate, Fast Foods and Sweeteners: Consumption and Health*, ed. Marlene R. Bishop, Nova Science Publishers, New York, 2010, pp. 245–255.

Beckett, Stephen T., "The History of Chocolate", in *The Science of Chocolate*, Royal Society of Chemistry, Cambridge, England, 2000, pp. 1–10.

Beckett, Stephen T., *The Science of Chocolate*, Royal Society of Chemistry, Cambridge, England, 2000.

Belluck, Pam, "Giving Alzheimer's patients their way, even chocolate", *The New York Times*, December 31, 2010, http://www.nytimes.com/2011/01/01/health/01care.html?_r=1 &pagewanted=print, accessed 14 June 2011.

Benedict, A. L., *Practical Dietetics*, G. P. Engelhard and Co., Chicago, 1904.

Benton, David, "The Biology and Psychology of Chocolate Craving", in *Coffee, Tea, Chocolate, and the Brain*, ed. Astrid Nehlig, CRC Press, Boca Raton, FL, 2004, pp. 205–218.

Berman, B. M., J. P. Swyers, S. M. Hartnoll, *et al.*, "The public debate over alternative medicine: The importance of finding a middle ground", *Alternative Therapies in Health and Medicine*, 2000, **6**, pp. 98–101.

Binder, Devin K., "The medical history of chocolate", *The Pharos*, 2001, **64**, pp. 22–26.

Biziere, Jean Maurice, "Hot beverages and the enterprising spirit in 18th-century Europe", *The Journal of Psychohistory*, 1979, **7**, pp. 135–145.

Blégny, Nicholas de, *Le Bon Usage du Thé, du Caffé, et du Chocolat pour la Préservation et pour la Guérison des Maladies*, Thomas Amaulry, Lyon, 1687.

Bletter, Nathaniel, and Douglas C. Daly, "Cacao and Its Relatives in South America", in *Chocolate in Mesoamerica: A Cultural History of Cacao*, ed. Cameron L. McNeil, University Press of Florida, Gainesville, 2006, pp. 31–68.

Bloom, Carole, *All About Chocolate: The Ultimate Resource to the World's Favorite Food*, Macmillan, New York, 1998.

Bontekoe, Cornelis, *Tractaat van het Excellenste Kruyd Thee: 't Welk Vertoond het Regte Gebruyk, en de Grote Kragten van 't Selve in Gesondheid, en Siekten: Benevens een Kort Kiscours op het Leven, de Siekte, en de Dood: Mitsgaders op de Medicijne, en de Medicijns van dese Tijd, en Speciaal van ons Land. Ten Dienste van die Gene, die lust Hebben, om Langer, Gesonder, en Wijser te Leven, als de Meeste Menschen nu in 't Gemeen Doen.* Pieter Hagen, 'sGravenhage, 1678.

Borchers, Andrea, Carl L. Keen, Sandra M. Hannum, *et al.*, "Cocoa and chocolate: Composition, bioavailability, and health implications", *Journal of Medicinal Food*, 2000, **3**, pp. 77–105.

Bordas, F., *De l'addition de Carbonate de Potassium aux Cacaos*, [n.p.], Paris, 1910.

Bordeaux B., L. R. Yanek, T. F. Moy, *et al.*, "Casual chocolate consumption and inhibition of platelet function", *Preventive Cardiology*, 2007, **10**, pp. 175–180.

Borg, Axel, and Adam Siegel, "Early Works on Chocolate: A Checklist", in *Chocolate: History, Culture, and Heritage*, eds. Louis Evan Grivetti and Howard Yana Shapiro, John Wiley & Sons, Hoboken, N. J., 2009, pp. 929–942.

Bourgaux, Albert, *Quatres Siècles d'histoire du Cacao et du Chocolat*, Office International du Cacao et du Chocolat, Brussels, 1935.

Boynton, Sandra, *Chocolate: The Consuming Passion*, Workman Publishing, New York, 1982.

Brenner, Joël Glenn, *The Emperors of Chocolate: Inside the Secret World of Hershey and Mars*, Random House, New York, 1999.

Brillat-Savarin, Jean Anthelme, *M. F. K. Fisher's Translation of The Physiology of Taste or Meditations on Transcendental Gastronomy*, North Point Press, San Francisco, 1986.

Brindle, Laura Pallas, and Bradley Foliart Olsen, "Adulteration: The Dark World of 'Dirty' Chocolate", in *Chocolate: History, Culture, and Heritage*, eds. Louis Evan Grivetti and Howard Yana Shapiro, John Wiley & Sons, Hoboken, N. J., 2009, pp. 625–634.

Brindle, Laura Pallas, and Bradley Foliart Olsen, "Digging for Chocolate in Charleston and Savannah", in *Chocolate: History, Culture, and Heritage*, eds. Louis Evan Grivetti and Howard Yana Shapiro, John Wiley & Sons, Hoboken, N. J., 2009, pp. 699–714.

Broekel, Ray, *The Chocolate Chronicles*, Wallace-Homestead Book Co., Lombard, IL, 1985.

Broekel, Ray, *The Great American Candy Bar Book*, Houghton Mifflin Co., Boston, 1982.

Brooks, E. St John, *Sir Hans Sloane: The Great Collector and his Circle*, Batchworth Press, London, 1954.

Brown, Peter B., *In Praise of Hot Liquors: The Study of Chocolate, Coffee and Tea-Drinking 1600–1850*, York Civic Trust, York, England, 1995.

Bruinsma, Kristen, and Douglas L. Taren, "Chocolate: Food or drug?", *Journal of the American Dietetic Association*, 1999, **99**, pp. 1249–1256.

Buijsse, B., E. J. Feskens, F. J. Kok, *et al.*, "Cocoa intake, blood pressure, and cardiovascular mortality: The Zutphen elderly study", *Archives of Internal Medicine*, 2006, **166**, pp. 411–417.

Buijsse, B., C. Weikert, D. Drogan, *et al.*, "Chocolate consumption in relation to blood pressure and risk of cardiovascular disease in German adults", *European Heart Journal*, 2010, **31**, pp. 1616–1623. Epub 2010 Mar 30.

Buitrago-Lopez, A., J. Sanderson, L. Johnson, *et al.*, "Chocolate consumption and cardiometabolic disorders: Systematic review and meta-analysis", *BMJ*, 2011, **343**, d4488.

Burnby, J. G. L., "Pharmacy and the cocoa bean", *The Pharmaceutical Historian*, 1984, **14**, pp. 9–12.

Busenberg, Bonnie, *Vanilla, Chocolate & Strawberry: The Story of Your Favorite Flavors*, Lerner Publications, Minneapolis, MN, 1994.

Caligiani, Augusta, Martina Cirlini, and Gerardo Palla, "Cocoa (*Theobroma Cacao L.*) Catechins: Occurrence, Health Effects and Modifications during Processing", in *Chocolate, Fast Foods and Sweeteners: Consumption and Health*, ed. Marlene R. Bishop, Nova Science Publishers, New York, 2010, pp. 231–244.

Carletti, Francesco, *My Voyage Around the World, Ragionamenti di Francesco Carletti Fiorentino sopra le cose da lui vedute ne' suoi viaggi*, translated from the Italian by Herbert Weinstock, Pantheon Books, New York, 1964.

Cassel, E. J., "The nature of suffering and the goals of medicine", *New England Journal of Medicine*, 1982, **306**, pp. 639–645.

Chamberlayne, John, *The Natural History of Coffee, Thee, Chocolate, Tobacco, in four several Sections; with a Tract of Elder and Juniper-Berries, Shewing how Useful they may be in our Coffee-Houses: And also the way of making Mum, With some Remarks upon that Liquor. Collected from the Writings of the best Physicians, and Modern Travellers*, Christopher Wilkinson, London, 1682.

Chevallier, J. B. A., *Hygiène Alimentaire: Mémoire sur le Chocolat*, Paris, 1871.

Chiva, Matty, "Cultural and Psychological Approaches to the Consumption of Chocolate", in *Chocolate and Cocoa: Health and Nutrition*, ed. Ian Knight, Blackwell Science, Oxford, England, 1999, pp. 321–338.

"Chocolate and cocoa", *Scientific American*, 1918, **86**, p. 232.

Chocolate Manufacturers Association of the U.S.A., *The Story of Chocolate*, Chocolate Manufacturers Association of the U.S.A., [n.p.], 1960.

Chocolate Research Portal, https://cocoaknow.ucdavis.edu/ChocolateResearch, accessed 29 April 2012.

Churchman, A., "Chocolate and cocoa products", *The Pharmaceutical Journal*, 1935, **135**, pp. 134–135.

Clarence-Smith, William Gervase, *Cocoa and Chocolate, 1765–1914*, Routledge, London and New York, 2000.

Clarence-Smith, William Gervase, ed., *Cocoa Pioneer Fronts since 1800: The Role of Smallholders, Planters and Merchants*, Macmillan Press, Houndmills and London, 1996.

Clarke, W. Tresper, "The Literature of Cacao", in *Advances in Chemistry*, American Chemical Society, 1954, **10**, pp. 286–296.

Clusius, Carolus, *Aliquot Notae in Garciae Aromatum Historiam, Eiusdem Descriptiones Nonnullarum Stirpium, & Aliarum Exoticarum Rerum, quae a Generoso viro Francisco Drake Equite Anglo, & his Observatae sunt, qui eum in Longa illa Navigatione, qua Proximis Annis Universum Orbem Circumivit, Comitati sunt: & Quorundam Peregrinorum Fructuum quos Londini ab Amicis Accepit*, Christophor Plantin, Antwerp, 1582.

Coady, Chantal, *The Chocolate Companion: A Connoisseur's Guide to the World's Finest Chocolates*, Simon & Schuster, New York, 1995.

Coady, Chantal, *Chocolate: The Food of the Gods*, Chronicle Books, San Francisco, 1993.

"Cocoa and its adulterations", *Lancet*, 1851, **1**, pp. 552, 608, 631.

Coe, Sophie D., and Michael D. Coe, *The True History of Chocolate*, Thames and Hudson, London, 1996.

Coe, Sophie D., and Michael D. Coe, *La Verdadera Historia del Chocolate*, FCE, Mexico City, 1999.

Colmenero de Ledesma, A., *Chocolate; or, An Indian Drink*, J. Dakins, London, 1652.

Cook, James J., *Chewing Gum, Candy Bars, and Beer: The Army PX in World War II*, University of Missouri Press, Columbia, MO, 2009.

Cook, L. Russell, *Chocolate Production and Use*, Magazines for Industry, New York, 1963.

Cooper, Karen A., Jennifer L. Donovan, Andrew L. Waterhouse, *et al.*, "Cocoa and health: A decade of research", *British Journal of Nutrition*, 2008, **99**, pp. 1–11.

Cooper, Thomas, *A Treatise of Domestic Medicine to Which is Added a Practical System of Domestic Cookery*, G. Getz, Reading, PA, 1824.

Corti, Roberto, Andreas J. Flammer, Norman K. Hollenberg, *et al.*, "Cocoa and cardiovascular health", *Circulation*, 2009, **119**, pp. 1433–1441.

Cox, Cat, *Chocolate Unwrapped: The Politics of Pleasure*, Women's Environmental Network, London, 1993.

Dakin, Karen, and Søren Wichmann, "Cacao and chocolate: A Uto-Aztecan perspective", *Ancient Mesoamerica*, 2000, **11**, pp. 55–75.

Dand, Robin, *The International Cocoa Trade*, 2nd edition, CRC Press, Boca Raton, FL; Woodhead Pub., Cambridge, England, 1999.

Dear, P., "*Totius in verba*: Rhetoric and authority in the early Royal Society", *Isis*, 1985, **76**, pp. 145–161.

Demarco, Giuseppe, *Josephi Demarco Medici Melitensis Philosophiæ, & Universitatis Monspelliensis medicinæ doctoris De Lana Ritè in Secunda, & Adversa valetudine adhibenda: Opus, quo Villosæ Vestis nudi contactus Præstantia, &c Actio Staticæ experimentis perspicuè, Utilitates fusè demonstrantur, Noxæ diligenter expenduntur. Adjecta est ad calcem Dissertatio de usu, et abusu chocolatæ in Re Medica, & Morali*, Melitæ: in Palatio, & ex Typographia C. S. S. Apud D. Nicolaum Capacium ejus Typographum, 1759.

De Quélus, D., *The Natural History of Chocolate; Being a Distinct and Particular Account of the Cocoa-tree, its Growth and Culture, and the Preparation, Excellent Properties, and Medicinal Vertues of its Fruit: Wherein the Errors of those who have wrote upon this Subject are discover'd; the Best Way of Making Chocolate is Explained; and Several Uncommon Medicines Drawn From It, are communicated*, 2nd edition, translated by Richard Brookes, J. Roberts, London, 1730.

Dillinger, Teresa, Patricia Barriga, Sylvia Escárcega, *et al.*, "Food of the Gods: Cure for humanity? A cultural history of the medicinal and ritual use of chocolate", *JN The Journal of Nutrition*, 2000, **130 Supplement**, pp. 2057S–2072S.

Djoussé, L., P. N. Hopkins, D. K. Arnett, *et al.*, "Chocolate consumption is inversely associated with calcified atherosclerotic plaque in the coronary arteries: The NHLBI Family Heart Study", *Clinical Nutrition*, 2011 Feb, **30**, pp. 38–43. Epub 2010 Jul 22.

Djoussé, L., P. N. Hopkins, K. E. North, *et al.*, "Chocolate consumption is inversely associated with prevalent coronary heart disease: The National Heart, Lung, and Blood Institute Family Heart Study", *Clinical Nutrition*, 2011 Apr, **30** pp. 182–187. Epub 2010 Sep 19.

Dodge, Bertha S., *Plants that Changed the World*, Little, Brown & Company, Boston & Toronto, 1959.

Doutre-Roussel, Chloé, *The Chocolate Connoisseur*, Penguin, New York, 2006.

Dreiss, Meredith L., and Sharon Edgar Greenhill, *Chocolate: Pathway to the Gods*, University of Arizona Press, Tucson, 2008.

Dufour, Philippe Sylvestre, *De l'Usage du Caphé, du Thé et du Chocolat*, Jean Girin & Barthélemy Rivière, Lyon, 1671.

Dufour, Philippe Sylvestre, *Traitez Nouveaux et Curieux du Café, du Thé et du Chocolat*, J.B. Deville, Lyon, 1688.

Dufour, P. S., *The Manner of Making Coffee, Tea and Chocolate as it is Used in Most Parts of Europe, Asia, Africa and America, With their Vertues*, W. Crook, London, 1685.

Duke, J. A., *CRC Handbook of Medicinal Herbs*, CRC Press, Boca Raton, FL, 1985.

Duke, James A., *Isthmian Ethnobotanical Dictionary*, 3rd edition, Scientific Publishers, Jodphur, India, 1986.

Duncan, Daniel, *Wholesome Advice Against the Abuse of Hot Liquors, Particularly of Coffee, Chocolate, Tea, Brandy, and Strong-Waters*, M. Rhodes and A. Bell, London, 1706.

Elliott, Catherine S., "Curing Irrationality with Chocolate Addiction", in *Chocolate: Food of the Gods*, ed. Alex Szogyi, Greenwood Press for Hofstra University, Westport, CT, 1997, pp. 19–34.

Ellis, Aytoun, *The Penny Universities : A History of the Coffee-Houses*, Secker & Warburg, London, 1956.

Ellis, Markman, ed., *Eighteenth-Century Coffee-House Culture*, Pickering & Chatto, London, 2006.

Eskes, A. B., and Y. Efron, eds., *Global Approaches to Cocoa Germplasm Utilization and Conservation*, CFC, Amsterdam, The Netherlands, 2006.

Faust, Betty Bernice, "Cacao beans and chili peppers: Gender socialization in the cosmology of a Yucatec Maya curing ceremony", *Sex Roles*, 1998, **39**, pp. 603–641.

Feinstein, Alvan R., *Clinical Judgment*, Williams and Wilkins, Baltimore, 1967.

Felici, Giovanni Batista, *Parere Intorno all'uso della Cioccolata: Scritto in una Lettera*, Appresso G. Manni, Florence, Italy, 1728.

Feltwell, John, assisted by Nathalie Bailleux, "Cacao Plantations", in *The Book of Chocolate*, Nathalie Bailleux, Hervé Bizeul, John Feltwell, *et al.*, Flammarion, Paris & New York, 1996, pp. 17–58.

Fernández-Murga, L., J. J. Tarín, M. A. García-Perez, *et al.*, "The impact of chocolate on cardiovascular health", *Maturitas*, 2011, **69**, pp. 312–321.

Few, Martha, "Chocolate, sex and disorderly women in late-seventeenth- and early-eighteenth-century Guatemala", *Ethnohistory*, 2005, **52**, pp. 674–687.

Fisher, Naomi D. L., and Norman K. Hollenberg, "Flavanols for cardiovascular health: The science behind the sweetness", *Journal of Hypertension*, 2005, **23**, pp. 1453–1459.

Flammer, Andreas J., Frank Hermann, Isabella Sudano, *et al.*, "Dark chocolate improves coronary vasomotion and reduces platelet reactivity", *Circulation*, 2007, **116**, pp. 2376–2382.

Fordyce, G., "An Attempt to Improve the Evidence of Medicine", *Transactions of the Society for the Improvement of Medical and Chirurgical Knowledge*, 1793, **1**, p. 243.

Foster, George M., *Hippocrates' Latin American Legacy: Humoral Medicine in the New World*, Gordon and Breach, Langhorne, PA, 1994.

Francis, S. T., K. Head, P. G. Morris, *et al.*, "The effect of flavanol-rich cocoa on the fMRI response to a cognitive task in healthy young people", *Journal of Cardiovascular Pharmacology*, 2006, **47**, pp. S215–S220.

Franke, Erwin, *Kakao, Tee und Gewürze*, A. Hartleben's Verlag, Wien & Leipzig, 1914.

Franklin, Alfred, *Le Café, le Thé & le Chocolat*, Series La Vie Privée d'autrefois, vol. 13, E. Plon, Nourriet et cie., Paris, 1893.

Fries, Joseph H., "Chocolate: A review of published reports of allergic and other deleterious effects, real or presumed", *Annals of Allergy*, 1978, **41**, pp. 195–207.

Fuller, Linda K., *Chocolate Fads, Folklore & Fantasies: 1,000+ Chunks of Chocolate Information*, Haworth Press, New York, 1994.

Gage, Thomas, *The English-American: A New Survey of the West Indies, 1648*, ed. A. P. Newton, George Routledge & Sons, London, 1928.

Galvin, Ruth Mehrtens, "Sybaritic to some, sinful to others, but how sweet it is!", *Smithsonian*, 1986, **16**, pp. 54–64.

Giuntini, Girolamo, *Altro Parere Intorno alla Natura ed all'uso della Cioccolata Disteso in Forma di Lettera Indirizzata all'illustrissimo Signor Conte Armando di Woltsfeitt*, [n.p.], Firenze, 1728 as cited in Francesco Merli, *Riflessioni Storico: Mediche Intorno All Uso Della Cioccolata*, [n.p.], Naples, 1779.

Golomb, Beatrice A., Sabrina Koperski, and Halbert L. White, "Association between more frequent chocolate consumption and lower body mass index", *Archives of Internal Medicine*, 2012, **172**, pp. 519–521.

"*Good Nutrition Makes Good Sense*", Hershey Foods Corporation Collection, Accession 87006, Box B-11, Folder 40, 1982.

Gott, Philip Porter, and L. F. Van Houten, *All About Candy and Chocolate: A Comprehensive Study of the Candy and Chocolate Industries*, National Confectioners' Association of the United States, Chicago, 1958.

Graziano, Martha Makra, "Food of the Gods as mortals' medicine: The uses of chocolate and cacao products", *Pharmacy in History*, 1998, **40**, pp. 132–146.

Green Grass Jingle Book for Little Folks, Hershey Foods Corporation Collection, Accession 87006, Box B-11, Folder 2, ca. 1915.

Grivetti, Louis Evan, "Chocolate and the Boston Smallpox Epidemic of 1764", in *Chocolate: History, Culture, and Heritage*, eds. Louis Evan Grivetti and Howard Yana Shapiro, John Wiley & Sons, Hoboken, N. J., 2009, pp. 89–98.

Grivetti, Louis Evan, "Dark Chocolate: Chocolate and Crime in North America and Elsewhere", in *Chocolate: History, Culture, and Heritage*, eds. Louis Evan Grivetti and Howard Yana Shapiro, John Wiley & Sons, Hoboken, N. J., 2009, pp. 255–262.

Grivetti, Louis Evan, "Medicinal Chocolate in New Spain, Western Europe, and North America", in *Chocolate: History, Culture, and Heritage*, eds. Louis Evan Grivetti and Howard Yana Shapiro, John Wiley & Sons, Hoboken, N. J., 2009, pp. 67–88.

Grivetti, Louis Evan, and Howard Yana Shapiro, eds., *Chocolate: History, Culture, and Heritage*, John Wiley & Sons, Hoboken, N. J., 2009.

Groopman, Jerome, *How Doctors Think*, Houghton Mifflin Company, New York, 2007.

Guerra, Francisco, "Aztec medicine", *Medical History*, 1966, **10**, pp. 315–388.

Guerra, Francisco, "Maya medicine", *Medical History*, 1964, **8**, pp. 31–43.

Guiltinan, Mark J., "Genomics of *Theobroma cacao*, 'the Food of the Gods' ", in *Genomics of Tropical Crop Plants*, eds. P. H. Moore and R. Ming, Springer, New York, 2008, pp. 145–170.

Gupta, M. P., P. N. Solís, A. I. Calderón, *et al.*, "Medical ethnobotany of the Teribes of Bocas del Toro, Panama", *Journal of Ethnopharmacology*, 2005, **96**, pp. 389–401.

Guthrie, G. J., *On Gun-Shot Wounds of the Extremities*, Longman, Hurst, Rees, Orme and Browne, London, 1815.

Hall, Grant D., Stanley M. Tarka, Jr., W. Jeffrey Hurst, *et al.*, "Cacao residues in ancient Maya vessels from Rio Azul, Guatemala", *American Antiquity*, 1990, **55**, pp. 138–143.

Hamed, Miruais S., Steven Gambert, Kevin P. Bliden, *et al.*, "Dark chocolate effect on platelet activity, C-reactive protein and lipid profile: A pilot study", *Southern Medical Journal*, 2008, **101**, pp. 1203–1208.

Harnack, E., "Zur streitfrage über den fettgehalt in den handelssorten des kakaos", *Deutsche Medizinische Wochenschrift*, 1906, **32**, pp. 1041–1043.

Harwich, Nikita, *Histoire du Chocolat*, Editions Desjonquères, Paris, 1992.

Hassall, Arthur Hill, *Food and its Adulterations; Comprising the Reports of the Analytical Sanitary Commission of "The Lancet" for the Years 1851 to 1854 Inclusive*, Longman, Brown, Green and Longmans, London, 1855.

Hawkins, F. B., *Elements of Medical Statistics*, Longman, London, 1829.

Head, Brandon, *The Food of the Gods: A Popular Account of Cocoa*, R. Brimley Johnson, London, 1903.

Heiss, Christian, Sarah Jahn, Melanie Taylor, *et al.*, "Improvement of endothelial function with dietary flavanols is associated with mobilization of circulating angiogenic cells in patients with coronary artery disease", *Journal of the American College of Cardiology*, 2010, **56**, pp. 218–224.

"*Hershey, Color It with Happy*", Hershey Foods Corporation Collection, Accession 87006, Box B-11, Folder 32, ca. 1968–1977.

[Hershey Community Archives], "Hershey's chocolate and the war effort", *Call to Duty*, 2011, **6**, p. 8.

"*Hershey Milk Chocolate: More Sustaining Than Meat*", Hershey Food Corporation Collection, Accession 87006, Box B-11, Folder 1, ca. 1904.

Hertog, Michaël G. L., Daan Kromhout, Christ Aravanis, *et al.*, "Flavonoid intake and long-term risk of coronary heart disease and cancer in the Seven Countries Study", *Archives of Internal Medicine*, 1995, **155**, pp. 381–386.

Hetherington, Marion M., and Jennifer I. Macdiarmid, "'Chocolate addiction': A preliminary study of its description and its relationship to problem eating", *Appetite*, 1993, **21**, pp. 233–246.

Higgs, Catherine, *Chocolate Islands: Cocoa, Slavery, and Colonial Africa*, Ohio University Press, Athens, OH, 2012.

Hinkle, S. F., *Fuel Values of Foods*, Hershey Foods Corporation Collection, Accession 87006, Box B-11, Folder 36, ca. 1936–1949.

"Historicus" [Richard Cadbury], *Cocoa: All About It*, S. Low, Marston & Co., London, 1892.

Hodgson, Jonathan M., Amanda Devine, Valerie Burke, *et al.*, "Chocolate consumption and bone density in older women", *American Journal of Clinical Nutrition*, 2008, **87**, pp. 175–180.

Hollenberg, N. K., Gregorio Martinez, Marji McCullough, *et al.*, "Aging, acculturation, salt intake, and hypertension in the Kuna of Panama", *Hypertension*, 1997, **29**, pp. 171–176.

Hollenberg, N. K., "Chocolate: God's gift to mankind? Maybe!", 2005, http://www.earthtimes.org/article/news/3849.html, accessed 31 May 2011.

http://www.ed.psu.edu/icik/, accessed 30 April 2012.

Hudson, R. P., "Polypharmacy in twentieth century America", *Clinical Pharmacology & Therapeutics*, 1968, **9**, pp. 2–10.

Hughes, William, *The American Physitian or A Treatise of the Roots, Plants, Trees, Shrubs, Fruit, Herbs &c. Growing in the English Plantations in America: Describing the Place, Time, Names, Kindes, Temperature, Vertues and Uses of them, either for Diet, Physick, &c. Whereunto is added A Discourse of the Cacao-nut Tree, and the use of its Fruit; with all the ways of making of Chocolate. The like never extant before*, J. C. for William Crook, London, 1672.

Hurst, W.J., "The Determination of Cacao in Samples of Archaeological Interest", in *Chocolate in Mesoamerica: A Cultural History of Cacao*, ed. Cameron L. McNeil, University Press of Florida, Gainesville, 2006, pp. 105–113.

Hurst, W. J., "A review of HPLC methods for the determination of selected biogenic amines in foods", *Journal of Liquid Chromatography*, 1990, **13**, pp. 1–23.

Hurst, W. Jeffrey, Robert A. Martin, Jr., Stanley M. Tarka, Jr., Grant D. Hall, *et al.*, "Authentication of cocoa in Maya vessels using high-performance liquid chromatographic techniques", *Journal of Chromatography*, 1989, **466**, pp. 279–289.

Hurst, W. J., R. A. Martin, B. L. Zoumas, *et al.*, "Biogenic amines in chocolate", *Nutrition Reports International*, 1982, **26**, pp. 1081–1086.

Hurst, W. Jeffrey, Stanley M. Tarka, Jr., Terry G. Powis, *et al.*, "Archaeology: Cacao usage by the earliest Maya civilization", *Nature*, 2002, **418**, p. 289–290.

Hurst, W.J., and P.B. Toomey, "High performance liquid chromatographic determination of four biogenic amines in chocolate", *Analyst*, 1981, **106**, pp. 394–404.

"Hysteria and chocolate", *Confectioners' Journal*, 1906, **32**, p. 69.

"In praise of chocolate", *Confectioners' Journal*, 1907, **33**, p. 94.

"The increasing use of cocoa and chocolate in America", *Confectioners' Journal*, 1906, **32**, p. 72.

Jacobsen, Rowan, *Chocolate Unwrapped: The Surprising Health Benefits of America's Favorite Passion*, Invisible Cities Press, Montpelier, VT, 2003.

Jameson, Eric, *The Natural History of Quackery*, Michael Joseph, London, 1961.

Joyce, Rosemary A., and John S. Henderson, "From feasting to cuisine: Implications of archaeological research in an early Honduran village", *American Anthropologist*, 2007, **109**, pp. 642–653.

Jumelle, Henri, *Le Cacaoyer: Sa Culture et son Exploitation dans tous les Pays de Production*, Augustin Challamel, Paris, 1900.

Katz, D. L., K. Doughty, and A. Ali, "Cocoa and chocolate in human health a disease", *Antioxidants & Redox Signaling*, 2011, **15**, pp. 2779–2811.

Katz, David L., "Health Effects of Chocolate", in *Nutrition in Clinical Practice: A Comprehensive Evidence-Based Manual for the Practitioner*, 2nd edition, David L. Katz, Lippincott Williams & Wilkins, Philadelphia, 2008, pp. 391–396.

Kean, B. H. "The blood pressure of the Cuna indians", *American Journal of Tropical Medicine and Hygiene*, 1944, **24**, pp. 341–343.

Keen, Carl L., "Chocolate: Food as medicine/medicine as food", *Journal of the American College of Nutrition*, 2001, **20**, pp. 436S–439S.

Kelly, Christopher, "Chocolate and North American Whaling Voyages", in *Chocolate: History, Culture, and Heritage*, eds. Louis Evan Grivetti and Howard Yana Shapiro, John Wiley & Sons, Hoboken, N. J., 2009, pp. 413–424.

Khodorowsky, Katherine, and Hervé Robert, *The Little Book of Chocolate*, Flammarion, Luzon, France, 2001.

Knapp, Arthur W., *Cocoa and Chocolate: Their History from Plantation to Consumer*, Chapman and Hall, London, 1920.

Knekt, Paul, Jorma Kumpulainen, Ritva Järvinen, *et al.*, Flavonoid intake and risk of chronic diseases", *American Journal of Clinical Nutrition*, 2002, **76**, pp. 560–568.

Knight, Ian, ed., *Chocolate and Cocoa: Health and Nutrition*, Blackwell Science, Oxford, England, 1999.

Kolpas, Norman, *The Chocolate Lovers' Companion*, The Felix Gluck Press, Twickenham, England, 1977.

Kris-Etherton, P. M., J. A. Derr, V. A. Mustad, *et al.*, "Effects of a milk chocolate bar per day substituted for a high-carbohydrate snack in young men on an NCEP/AHA Step 1 Diet", *American Journal of Clinical Nutrition*, 1994, **60**, pp. 1037S–1042S.

Kunow, Marianna Appel, *Maya Medicine: Traditional Healing in Yucatán*, University of New Mexico Press, Albuquerque, 2003.

Kunst in Schokolade [Chocolate Art], Hatje Cantz, Ostfildern-Ruit, 2005.

Labane, Pierre, "The History of Chocolate", in *The Book of Chocolate*, Nathalie Bailleux, Hervé Bizeul, John Feltwell, *et al.*, Flammarion, Paris and New York, 1996, pp. 59–104.

Lacroix, Monique, "Polyphenols in Cocoa: Influence of Processes on Their Composition and Biological Activities", in *Chocolate, Fast Foods and Sweeteners: Consumption and Health*, ed. Marlene R. Bishop, Nova Science Publishers, New York, 2010, pp. 183–197.

Laine, Victoria, *Health by Chocolate: Radical New Recipes & Nutritional Know-How*, Owl Medicine Books, [n.p.], 2008.

Lang, C. Max, *The Impossible Dream: The Founding of the Milton S. Hershey Medical Center of the Pennsylvania State University*, AuthorHouse, Bloomington, IN, 2010.

Laudan, Rachel, and Jeffrey M. Pilcher, "Chiles, chocolate, and race in New Spain: Glancing backward to Spain or looking forward to Mexico?", *Eighteenth-Century Life*, 1999, **23**, pp. 59–70.

Lavedan, Antonio, *Tratado de Los Usos, Abusos, Propiendades y Virtudes del Tabaco, Café, Té y Chocolate*, Imprenta Real, Madrid, Spain, 1796.

Laws, Bill, *Fifty Plants that Changed the Course of History*, David & Charles, Cincinnati, OH, 2010.

Lebovitz, David, *The Great Book of Chocolate: The Chocolate Lover's Guide, with Recipes*, Ten Speed Press, Berkeley, 2004.

LeCount, Lisa J., "Like water for chocolate: Feasting and political ritual among the Late Classic Maya at Xunantunich, Belize", *American Anthropologist*, 2001, **103**, pp. 935–953.

Lee, Ki Won, Nam Joo Kang, Min-Ho Oak, *et al.*, "Cocoa procyanidins inhibit expression and activation of MMP-2 in vascular smooth muscle cells by direct inhibition of MEK and MT1-MMP activities", *Cardiovascular Research*, 2008, **79**, pp. 34–41.

Lémery, M. L., *A Treatise of all Sorts of Foods, both Animal and Vegetable: Also of Drinkables: Giving an Account How to Chuse the Best Sort of all Kinds*, translated by D. Hay, W. Innys, T. Longman, and T. Shewell, London, 1745.

Levin, A., "The Cochrane collaboration", *New England Journal of Medicine,* 2001, **135**, pp. 309–312.

Liebowitz, Michael R., *The Chemistry of Love*, Little Brown, Boston, 1983.

Lind, J., *A Treatise on the Scurvy*, 3rd edition, Crowder, Wilson, Nicholls, Cadell, Becket, Pearch and Woodfall, London, 1772.

Lippi, Donatella, "Chocolate and medicine: Dangerous liaisons?", *Nutrition*, 2009, **25**, pp. 1100–1103.

Long, Eula, *Chocolate: From Mayan to Modern*, Aladdin Books, New York, 1950.

Lopez, Ruth, *Chocolate: The Nature of Indulgence*, H. N. Abrams in association with the Field Museum, New York, 2002.

Louis, P.-C.-A., *Researches on the Effects of Bloodletting in Some Inflammatory Diseases, and On the Influence of Tartarized Antimony and Vessication in Pneumonitis*, translated by C.G. Putnam, Hillard, Gray, Boston, 1836.

"*M&M'S® Chocolate Candies Go Green Just In Time For Valentine's Day*", http://multivu.prnewswire.com/mnr/mars/31278/, accessed 28 April 2012.

Macfadden, Bernarr, *Hershey, The Chocolate Town*, Hershey Chocolate Company, Hershey, PA, ca. 1923.

MacGregor, Arthur, ed., *Sir Hans Sloane: Collector, Scientist, Antiquary, Founding Father of the British Museum*, British Museum Press, London, 1994.

MacLeod, Murdo J., "Cacao", in *The Cambridge World History of Food*, eds. Kenneth F. Kiple and Kriemhild Coneè Ornelas, Cambridge University Press, Cambridge, UK and New York, 2000, Vol. 1, pp. 635–641.

Mallet, D., *The Life of Francis Bacon, Lord Chancellor of England*, A. Millar, London, 1740.

Marcus, Adrianne, "A Brief History of Chocolate", in *The Chocolate Bible*, Adrianne Marcus, G. P. Putnam's Sons, New York, 1979, pp. 27–35.

Marcus, Adrianne, *The Chocolate Bible*, G. P. Putnam's Sons, New York, 1979.

Mariani, John F., "Sweet talk and chocolate", *MD*, 1993, **37**, pp. 89–91.

Marshall, Janette, *The Here's Health Alternative Chocolate Book: Over 100 Healthy Carob Recipes*, Century Press, London, 1986.

Martin, Paul, *Sex, Drugs & Chocolate: The Science of Pleasure*, Fourth Estate, London, 2008.

Massiallot, François, *Nouvelle Instruction pour les Confitures, les Liqueurs, et les Fruits*, Seconde édition, augmentée, [n.p.], Paris, 1734.

Matthews, J. R., *Quantification and the Quest for Medical Certainty*, Princeton University Press, Princeton, N. J., 1995.

McFadden, Christine, and Christine France, *Chocolate: Cooking with the World's Best Ingredient*, Hermes House, New York, 2001.

McGee, Harold, *On Food and Cooking: The Science and Lore of the Kitchen*, Charles Scribner's Sons, New York, 1984.

McIntosh, William Alex, Karen S. Kubena and Wendall A. Landmann, "Chocolate and Loneliness among the Elderly", in *Chocolate: Food of the Gods*, ed. Alex Szogyi, Greenwood Press for Hofstra University, Westport, CT, 1997, pp. 3–10.

McMahon, James D., Jr., *Built on Chocolate: The Story of the Hershey Chocolate Company*, General Publishing Group, Los Angeles, CA, 1998.

McNeil, Cameron L., ed., *Chocolate in Mesoamerica: A Cultural History of Cacao*, University Press of Florida, Gainesville, 2006.

McNeil, Cameron L., "Introduction: The Biology, Antiquity, and Modern Uses of the Chocolate Tree (*Theobroma cacao L.*)", in *Chocolate in Mesoamerica: A Cultural History of Cacao*, ed. Cameron L. McNeil, University Press of Florida, Gainesville, 2006, pp. 1–30.

Meisner, Leonhardus Ferdinandus, *De Caffe, Chocolate, Herbae Thee ac Nicotianae Natura, Usu, et Abusu Anacrisis Medico-Historico-Diaetetica*, J. F. Rüdiger, Nuremberg, 1721.

Messer, A'ndrea Elyse, "Cocoa (data) crunch", *Research Penn State*, 2011, **32**, p. 21.

Meyrick, William, *The New Family Herbal; or, Domestic Physician: Enumerating, with Accurate Descriptions, all the known vegetables which are any way remarkable for medical efficacy; with an account of their virtues in the several diseases incident to the human frame*, Thomas Pearson, Birmingham, England, 1790.

Millar, J., *Observations on the Prevailing Diseases in Great Britain*, 2nd edition, Millar, London, 1798.

Miller, Frederic P., Agnes F. Vandome, and John McBrewster, eds., *Health Effects of Chocolate: Chocolate, Epicureanism, Cocoa, Dark Chocolate, Circulatory System, Anticancer*, Alphascript Publishing, Beau Bassin, Mauritius, 2010.

Minifie, Bernard W., *Chocolate, Cocoa and Confectionery: Science and Technology*, 3rd edition, Van Nostrand Reinhold, New York, 1989.

Mink, Pamela J., Carolyn G. Scrafford, Leila M. Barraj, *et al.*, "Flavonoid intake and cardiovascular disease mortality: A prospective study in postmenopausal women", *American Journal of Clinical Nutrition*, 2007, **85**, pp. 895–909.

Minton, Phillip, *Chocolate: Healthfood of the Gods: Unwrap the Secrets of Chocolate for Health, Beauty and Longevity*, 2011, available through Amazon.com and bn.com.

Mitchell, Donald G., *The Chocolate Industry*, Bellman, Boston, 1951.

Molinari, Enrico, and Edward Callus, "Psychological Drivers of Chocolate Consumption", in *Chocolate and Health*, eds. Rodolfo Paoletti, Andrea Poli, Ario Conti, *et al.*, Springer Verlag Italia, Milan, 2012, pp. 137–146.

Monagas, Maria, Nasiruddin Khan, Cristina Andres-Lacueva, *et al.*, "Effect of cocoa powder on the modulation of inflammatory biomarkers in patients at high risk of cardiovascular disease", *American Journal of Clinical Nutrition*, 2009, **90**, pp. 1144–1150.

Montgomery, Kathryn, *How Doctors Think: Clinical Judgment and the Practice of Medicine*, Oxford University Press, Oxford, 2005.

Morand, Jean François Clément, *Quaestio Medica ... An Senibus Chocolatae Potus?* [Quillau, Paris, 1749].

Moreau, René, *Du Chocolat: Discours Curieux Divise en Quatre Parties*, Sebastien Cramoisy, Paris, 1643.

Morgan, Jeff, "Chocolate: A flavor and texture unlike any other", *American Journal of Clinical Nutrition*, 1994, **60 Supplement**, pp. 1065S–1067S.

Mortimer, W. Golden, *History of Coca: "The Divine Plant" of the Incas*, Fitz Hugh Ludlow Memorial Library Edition, And/Or Press, San Francisco, 1974.

Morton, Marcia, *Chocolate: An Illustrated History*, Crown, New York, 1986.

Moss, Sarah, and Alexander Badenoch, *Chocolate: A Global History*, Reaktion Books, London, 2009.

Mossu, Guy, *Cocoa*, Macmillan Press, London, 1992.

Mostofsky, E., E. B. Levitan, A. Wolk, *et al.*, "Chocolate intake and incidence of heart failure: A population-based prospective study of middle-aged and elderly women", *Circulation. Heart Failure*, 2010, **3** pp. 612–616. Epub 2010 Aug 16.

Navier, Pierre Toussaint, *Bemerkungen über den Cacao und die Chocolate, worinnen der Nutzen und Schaden untersuchet wird, der aus dem Genusse dieser nahrhaften Dinge entsehen kann: Alles auf*

Erfahrung und zergliedernde Versuche mit der Cacao-Mandel gebauet; Nebst einigen Erinnerungen über das System des Hrn. De-La-Müre, betreffend das Schlagen der Puls-adern, Saalbach, Leipzieg, 1775.

Nestlé Company, *The History of Chocolate and Cocoa* [leaflet, n.p., n.d.].

Nestle, Marion, "The Role of Chocolate in the American Diet: Nutritional Perspectives", in *Chocolate: Food of the Gods*, ed. Alex Szogyi, Greenwood Press for Hofstra University, Westport, CT, 1997, pp. 111–124.

Nevinson, Henry Woodd, *A Modern Slavery*, Harper & Brothers, London and New York, 1906.

Norton, Marcy, *Sacred Gifts, A History of Profane Tobacco and Chocolate Pleasures in the Atlantic World*, Cornell University Press, Ithaca and London, 2008.

Norton, Marcy, "Tasting empire: Chocolate and the European internalization of Mesoamerican aesthetics", *American Historical Review*, 2006, **111**, pp. 660–691.

Off, Carol, *Bitter Chocolate: The Dark Side of the World's Most Seductive Sweet*, The New Press, New York and London, 2008.

Ogata, Nisao, Arturo Gómez-Pompa, and Karl A. Taube, "The Domestication and Distribution of *Theobroma cacao L.* in the Neotropics", in *Chocolate in Mesoamerica: A Cultural History of Cacao*, ed. Cameron L. McNeil, University Press of Florida, Gainesville, 2006, pp. 69–89.

Othick, J. "The Cocoa and Chocolate Industry in the Nineteenth Century", in *The Making of the Modern British Diet*, eds. Derek Oddy and Derek Miller, Croom Helm, London and Rowman and Littlefield, Totowa, N.J., 1976, pp. 77–90.

Paoletti, Rodolfo, Andrea Poli, Ario Conti, and Francesco Visioli, eds., *Chocolate and Health*, Springer Verlag Italia, Milan, 2012.

Parker, Gordon, I. Parker, and H. Brotchie, "Mood state effects of chocolate", *Journal of Affective Disorders*, 2006, **92**, pp. 149–159.

Parker, James N., and Philip Parker, eds., *Chocolate: A Medical Dictionary, Bibliography and Annotated Research Guide to Internet References*, ICON Health Publications, San Diego, CA, 2003, http://www.netLibrary.com/urlapi.asp?action = summary&v = &bookid = 99889, accessed 28 April 2012.

Pavy, Frederick William, *Treatise on Food and Dietetics: Physiologically and Therapeutically Considered*, Henry C. Lea, Philadelphia, 1874.

Pearson, Debra A., Teresa G. Paglieroni, Dietrich Rein, *et al.*, "The effects of flavanol-rich cocoa and aspirin on ex vivo platelet function", *Thrombosis Research*, 2002, **106**, pp. 191–197.

Pech, Julie, *The Chocolate Therapist: Chocolate Remedies for a World of Ailments*, Trafford, Victoria, B. C., 2005.

Pech, Julie, *The Chocolate Therapist: A User's Guide to the Extraordinary Health Benefits of Chocolate*, Trafford, Victoria, B. C., 2006.

Pelchat, Marcia L., and Gary K. Beauchamp, "Sensory and Taste Preferences of Chocolate", in *Chocolate and Cocoa: Health and Nutrition*, ed. Ian Knight, Blackwell Science, Oxford, England, 1999, pp. 310–320.

Pelletier, Eugene, and Auguste Pelletier, *Le Thé et le Chocolat dans l'alimentation Publique aux Points de Vue Historique, Botanique, Physiologique, Hygiénique, Économique, Industriel, et Commercial*, Le Compagnie Française des Chocolats et des Thés, Paris, 1861.

Pepys, S., *Diary*, 1661, Wednesday 24 April, http://www.pepysdiary.com/archive/1661/04, accessed 31 May 2011.

Pérez-Cano, Francisco J., Teresa Pérez-Berezo, Sara Ramos-Romero, *et al.*, "Is There an Anti-Inflammatory Potential Beyond the Antioxidant Power of Cocoa?", in *Chocolate, Fast Foods and Sweeteners: Consumption and Health*, ed. Marlene R. Bishop, Nova Science Publishers, New York, 2010, pp. 85–104.

Pollan, Michael, *The Botany of Desire: A Plant's Eye View of the World*, Random House, New York, 2001.

Pomet, Pierre, *A Compleat History of Drugs, Written in French by Monsieur Pomet, Chief Druggist to the late French King Lewis XIV. To which is added what is further observable on the same subject, from Mess. Lemery and Tournefort, divided into three classes, Vegetable, Animal and Mineral; with their use in Physick, Chymistry, Pharmacy, and several other Arts*, 3rd edition, J. and J. Bonwicke, R. Wilkin [etc.], London, 1737.

Pomet, P., *Histoire Générale des Drogues, Simples et Composes*, J.-B. Loyson, A. Pillon et E. Ducastin, Paris, 1694.

Porter, Roy, *Health for Sale: Quackery in England, 1660–1850*, Manchester University Press, Manchester, 1989.

Powis, T. G., W. J. Hurst, M. del C. Rodríguez, *et al.*, "Oldest chocolate in the New World", *Antiquity*, 2007, **81**, pp. 302–305.

Presilla, Maricel E., *The New Taste of Chocolate: A Cultural and Natural History of Cacao with Recipes*, Ten Speed Press, Berkeley and Toronto, 2001.

Prior, M. E., "Bacon's man of science", *Journal of the History of Ideas*, 1954, **15**, pp. 348–370.

Pucciarelli, Deanna, "Chocolate as Medicine: Imparting Dietary Advice and Moral Values Through 19th Century North American Cookbooks", in *Chocolate: History, Culture, and Heritage*, eds., Louis Evan Grivetti and Howard Yana Shapiro, John Wiley & Sons, Hoboken, N. J., 2009, pp. 115–126.

Pucciarelli, Deanna, and James Barrett, "Twenty-First Century Attitudes and Behaviors Regarding the Medicinal Use of Chocolate", in *Chocolate: History, Culture, and Heritage*, eds. Louis Evan Grivetti and Howard Yana Shapiro, John Wiley & Sons, Hoboken, N. J., 2009, pp. 653–666.

Pucciarelli, Deanna L., and Louis E. Grivetti, "The medicinal use of chocolate in early North America", *Molecular Nutrition & Food Research*, 2008, **52**, pp. 1215–1227.

Quinton, A., *Francis Bacon*, Oxford University Press, Oxford, 1980.

Rabutin-Chantal, M., *Letters choisies*, Garnier Frères, Paris, 1878. Letters 11.2.1671, 15.4.1671, 13.5.1671, 23.10.1671.

Raloff, Janet, "Chocolatc hcarts: Yummy and good medicine?", *Science News*, 2000, **157**, pp. 188–189.

Ramiro-Puig, Emma, and Margarida Castell, "Cocoa: Antioxidant and immunomodulator", *British Journal of Nutrition*, 2009, **101**, pp. 931–940.

Reed-Danahay, Deborah, "Champagne and chocolate: "Taste" and inversion in a French wedding ritual", *American Anthropologist*, 1996, **98**, pp. 750–761.

Reid, Larry D., "Delicious or Addictive?", in *Chocolate, Fast Foods and Sweeteners: Consumption and Health*, ed. Marlene R. Bishop, Nova Science Publishers, New York, 2010, pp. 313–317.

Rein, Dietrich, Teresa G. Paglieroni, Ted Wun, *et al.*, "Cocoa inhibits platelet activation and function", *The American Journal of Clinical Nutrition*, 2000, **72**, pp. 30–35.

Rinzler, Carol Ann, *The Book of Chocolate*, St. Martin's Press, New York, 1977.

Robertson, Emma, *Chocolate, Women and Empire: A Social and Cultural History*, Manchester University Press, Manchester, 2010.

Robertson, R., *Observations on the Jail, Hospital or Ship Fever*, Murray, London, 1783.

Rogers, Kara, *Out of Nature: Why Drugs from Plants Matter to the Future of Humanity*, University of Arizona Press, Tucson, 2012.

Rogers, T. B., *A Century of Progress, 1831–1931: Cadbury, Bournville*, [Hudson & Kearns, London, 1931].

Rosenblum, Mort, *Chocolate: A Bittersweet Saga of Dark and Light*, North Point Press, New York, 2005.

Rössner, S., "Chocolate – divine food, fattening junk or nutritious supplementation?", *European Journal of Clinical Nutrition*, 1997, **51**, pp. 341–345.

Rozin, Paul, E. Levine, and C. Stoess, "Chocolate craving and liking", *Appetite*, 1991, **17**, pp. 199–212.

Rusconi, M., and A. Conti, "*Theobroma cacao L.*, the food of the Gods: A scientific approach beyond myths and claims", *Pharmacological Research*, 2010, **61**, pp. 5–13.

Rusnock, Andrea A., "The Weight of Evidence and the Burden of Authority: Case Histories, Medical Statistics and Smallpox Inoculation", in *Medicine in the Enlightenment*, ed. R. Porter, Rodopi, Amsterdam, 1995, pp. 289–315.

Ryan, Órla, *Chocolate Nations: Living and Dying for Cocoa in West Africa*, Zed Books, London and New York, 2011.

Sackett, D. L., W. M. C. Rosenberg, J. A. M. Gray, *et al.*, "Evidence-based medicine: What it is and what it isn't", *BMJ*, 1996, **312**, pp. 71–72.

Saint-Arroman, Auguste, *Coffee, Tea & Chocolate; Their Influence upon the Health, the Intellect & the Moral Nature of Man*, T. Ward, Philadelphia, 1846.

Sammarco, Anthony M., *The Baker Chocolate Company: A Sweet History*, History Press, Charleston, SC, 2009.

Satre, Lowell J., *Chocolate on Trial: Slavery, Politics and the Ethics of Business*, Ohio University Press, Athens, OH, 2005.

Saunders, W. Clarke, "Adulteration of cocoa and chocolate", *Confectioners' Journal*, 1895, **21**, pp. 64–65.

Savi, Lidia, "Is There a Relationship between Chocolate Consumption and Headache?", in *Coffee, Tea, Chocolate, and the Brain*, ed. Astrid Nehlig, CRC Press, Boca Raton, FL, 2004, pp. 219–226.

Schewe, Tankred, Hartmut Kühn, and Helmut Sies, "Flavonoids of cocoa inhibit recombinant human 5-lipoxygenase", *JN: The Journal of Nutrition*, 2002, **132**, pp. 1825–1829.

Schneider, Robert, and Joyce Schneider, *The Cardiologist's Wife's Chocolate Too! Diet: No Sugar, No Fat, and Luscious*, BookSurge Publishing, Charleston, SC, 2007.

Schuier, Maximilian, Helmut Sies, Beate Illek, *et al.*, "Cocoa-related flavonoids inhibit CFTR-mediated chloride transport across T84 human colon epithelia", *Journal of Nutrition*, 2005, **135**, pp. 2320–2325.

Selmi, Carlo, Claudio A. Cocchi, Mario Lanfredini, *et al.*, "Chocolate at heart: The anti-inflammatory impact of cocoa flavanols", *Molecular Nutrition & Food Research*, 2008, **52**, pp. 1340–1348.

Selmi, Carlo, Tin K. Mao, Carl L. Keen, *et al.*, "The anti-inflammatory properties of cocoa flavanols", *Journal of Cardiovascular Pharmacology*, 2006, **47 Supplement 2**, pp. S163–S171.

Shapin, S., "Pump and circumstance: Robert Boyle's literary technology", *Social Studies of Science*, 1984, **14**, pp. 487–494.

Sherr, Jeremy, *The Homoeopathic Proving of Chocolate*, [B. Jain Publishers, n.p., 1993].

Shiina, Yumi, Nobusada Funabashi, Kwangho Lee, *et al.*, "Acute effect of oral flavonoid-rich dark chocolate intake on coronary circulation, as compared with non-flavonoid white chocolate, by transthoracic Doppler echocardiography in healthy adults", *International Journal of Cardiology*, 2009, **131**, pp. 424–429.

Shrime, Mark G., Scott R. Bauer, Anna C. McDonald, *et al.*, "Flavonoid-rich cocoa consumption affects multiple cardiovascular risk factors in a meta-analysis of short-term studies", *The Journal of Nutrition*, 2011, **141**, pp. 1982–1988.

Shultes, R., and R. Raffauf, *The Healing Forest*, Dioscorides Press, Portland, 1990.

Smaridge, Norah, *The World of Chocolate*, J. Messner, New York, 1969.

Smith, Andrew F., ed., *The Oxford Encyclopedia of Food and Drink in America*, Oxford University Press, Oxford, England, 2004.

Snavely, J. R., "The nutritive value of chocolate and cocoa", *The Confectioners' Journal*, 1925, **51**, pp. 96–97.

Spadafranca, A., C. Martinez Conesa, S. Sirini, *et al.*, "Effect of dark chocolate on plasma epicatechin levels, DNA resistance to oxidative stress and total antioxidant activity in healthy subjects", *British Journal of Nutrition*, 2010, **103**, pp. 1008–1014. Epub 2009 Nov 5.

Stage, Sarah, *Female Complaints: Lydia Pinkham and the Business of Women's Medicine*, W. W. Norton, New York, 1979.

The Story of Chocolate and Cocoa: With a Brief Description of Hershey, "The Chocolate and Cocoa Town" and Hershey "The Sugar Town", Hershey Chocolate Corp., Hershey, PA, 1926, 1934.

Strother, Edward, *Materia Medica: or, a New Description of the Virtues and Effects of all Drugs, or Simple Medicines now in Use*, Charles Rivington, London, 1727.

Stubbe, H., *The Indian Nectar, or a Discourse Concerning Chocolate wherein the Nature of the Cacao-nut ... is Examined ... the Ways of Compounding and Preparing Chocolate are Enquired into; its Effects, as to its Alimental and Venereal Quality, as well as Medicinal (Specially in Hypochondriacal Melancholy) are Fully Debated*, A. Crook, London, 1662.

Szogyi, Alex, ed., *Chocolate: Food of the Gods*, Greenwood Press for Hofstra University, Westport, CT, 1997.

Tachibana, Chris, "Dr. Chocolate", *The Scientist*, 2010, **24**, pp. 23–24.

Tannahill, Reay, *Food in History*, Eyre Methuen, London, 1973.

Taubert, Dirk, R. Roesen, C. Lehmann, *et al.*, "Effects of low habitual cocoa intake on blood pressure and bioactive nitric oxide", *JAMA*, 2007, **298**, pp. 49–60.

Theobald, Mary Miley, "A cup of hot chocolate, s'good for what ails ya", *Colonial Williamsburg*, 2012, **34**, pp. 46–52.

Thomas, L. A., *Active Life: For People 50-Plus On The Move*, 5 April 2007, p 4.

Thompson, J. Eric S., "Notes on the use of cacao in middle America", *Notes on Middle American Archaeology and Ethnology,* Carnegie Institution of Washington, Department of Archaeology, 1956, **128**, pp. 95–116.

Tröhler, U., *To Improve the Evidence of Medicine: The 18th Century British Origins of a Critical Approach*, Royal College of Physicians of Edinburgh, Edinburgh, 2001.

Turner, Daniel, *Art of Surgery*, 2 vol., Rivington, Lacy and Clarke, London, 1725.

Turner, Daniel, *Discourse Concerning Fevers*, Oake, London, 1727.

Ure, Andrew, *A Dictionary of Arts, Manufactures, and Mines: Containing a Clear Exposition of their Principles and Practice*, Longman, Orme, Brown, Green, & Longmans, London, 1839.

Urquhart, D. H., *Cocoa*, Longmans, Green & Co., London, 1955.

Usmani, Omar S., Maria G. Belvisi, Hema J. Patel, *et al.*, "Theobromine inhibits sensory nerve activation and cough", *The FASEB Journal*, 2005, **19**, pp. 231–233.

Van Hall, C. J. J., *Cacao*, MacMillan & Co., London, 1914, 2nd edition 1932.

Vélez, Santiago Londoño, *The Virtues and Delights of Chocolate*, Compañía Nacional de Chocolates, Medellín, Colombia, 2003.

Visioli, Francesco, H. Bernaert, R. Corti, *et al.*, "Chocolate, lifestyle, and health", *Critical Reviews in Food Science and Nutrition*, 2009, **49**, pp. 299–312.

Vives, I., "Del arroz, pastas, garbanzos y chocolates, considerados como elementos del regimen alimenticio de los enfermos militares", *Gaceta de Sanidad Militar*, 1882, **8**, p. 176.

Wagner, Gillian, *The Chocolate Conscience*, Chatto and Windus, London, 1987.

Wan, Ying, Joe A. Vinson, Terry D. Etherton, *et al.*, "Effects of cocoa powder and dark chocolate on LDL oxidative susceptibility and prostaglandin concentrations in humans", *American Journal of Clinical Nutrition*, 2001, **74**, pp. 596–602.

Wasicky, R., and C. Wimmer, "Eine neue methode des nachweises der schalen im kakao", *Zeitschrift für Untersuchung der Nahrungs-und Genussmittel*, 1915, **30**, pp. 25–27.

Waterhouse, Debra, *Why Women Need Chocolate: Eat What You Crave to Look Good & Feel Great*, Hyperion, New York, 1995.

Watson, Ronald Ross, Victor R. Preedy, and Sherma Zibadi, eds., *Chocolate in Health and Nutrition*, Humana Press, New York, 2013.

Weil, Andrew, and Winifred Rosen, *From Chocolate to Morphine: Everything You Need to Know about Mind-Altering Drugs*, Revised edition, Houghton Mifflin, Boston, 2004.

Weil, Andrew T., "Special Report: Is Chocolate Addictive?", in, *Medical & Health Annual 1985*, ed. Ellen Bernstein, Encyclopaedia Britannica Inc., Chicago, 1984, pp. 221–223.

West, John A., "A Brief History and Botany of Cacao", in *Chilies to Chocolate: Food the Americas Gave the World*, eds. Nelson Foster and Linda S. Cordell, University of Arizona Press, Tucson and London, 1992, pp. 105–121.

Whymper, Robert, *Cocoa and Chocolate: Their Chemistry and Manufacture*, J. & A. Churchill, London, 1912.

Williams, C. Trevor, *Chocolate and Confectionary*, Leonard Hill, London, 1950, 3rd edition, 1964.

Williams, Edith C., *A Bibliography of the Nutritive Value of Chocolate and Cocoa with Quotations and Summaries*, prepared for the Hershey Chocolate Company by The American Food Journal Institute, Hershey, PA, [1925].

Wilson, Bee, *Swindled: The Dark History of Food Fraud, from Poisoned Candy to Counterfeit Coffee*, Princeton University Press, Princeton, NJ, 2008.

Wilson, Philip K., "The art of medicine: Centuries of seeking chocolate's medicinal benefits", *Lancet*, 2010, **376**, pp. 158–159.

Wilson, Philip K., "Chocolate as Medicine: A Changing Framework of Evidence Throughout History", in *Chocolate and Health*, eds. Rodolfo Paoletti, Andrea Poli, Ario Conti, and Francesco Visioli, Springer-Verlag Italia, Milan, 2012, pp. 1–15.

Wilson, Philip K., "Origins of science", *National Forum* (Journal of the National Honor Society, Phi Kappa Phi), 1996, **76**, pp. 39–43. Also reprinted in SIRS Renaissance electronic database, 1996.

Wilson, Philip K., *Surgery, Skin & Syphilis: Daniel Turner's London (1667–1741)*, Wellcome Institute Series in the History of Medicine, Clio Medica **54**, Rodopi Press, Amsterdam and Atlanta, 1999.

Wilson, Philip K., "Weighing medical evidence on a historical scale", *Hektoen International: A Journal of Medical Humanities*, 2011, **3**, www.hektoneinternational.org/Weighing-medical-evidence.html.

Wilson, Philip K., "William Salmon", in *New Dictionary of National Biography*, ed. H. C. G. Matthew, Oxford University Press, Oxford, 2004, **48**, pp. 734–735.

Wood, G. A. R., and R. A. Lass, *Cocoa*, Longman, London & New York, 1985.

Wood, G. B., and F. Bache, eds., *Dispensatory of the United States*, Gregg and Elliot, Philadelphia, 1834.

Woskresensky, Alexander, "Über das theobromin", *Liebig's Annalen der Chemie und Pharmcie*, 1842, **41**, pp. 125–127.

Yochum, Laura, Lawrence H. Kushi, Katie Meyer, *et al.*, "Dietary flavonoid intake and risk of cardiovascular disease in postmenopausal women", *American Journal of Epidemiology*, 1999, **149**, pp. 943–949.

Young, A. M., *The Chocolate Tree: A Natural History of Cacao*, Smithsonian Institution Press, Washington, D.C., 1994; Revised and expanded edition, University Press of Florida, Gainesville, 2007.

Young, Gordon, "Chocolate: Food of the gods", *National Geographic*, 1984, **166**, pp. 664–687.

Zipperer, Paul, *The Manufacture of Chocolate*, 3rd English Edition, Spon and Chamberlain, New York, 1915.

Subject Index